CITIES SQUARED

MAKING URBAN DATA LEGIBLE   BY AFTER THE FLOOD

CONTENTS   FOREWORD   5
                                      INTRODUCTION   9

                                      PATTERNS   18
                                      SHAPES   20
                                      LIMITS   22
                                      VISUAL HOOKS   26

                                      LONDON   30
                                      SEATTLE   34
                                      COPENHAGEN   35
                                      MUMBAI   36
                                      DÜSSELDORF   38
                                      AMSTERDAM   39
                                      HELSINKI   40
                                      AUCKLAND   42
                                      HAVANA   44
                                      CAPE TOWN   45
                                      MILAN   46
                                      STOCKHOLM   47
                                      KAMPALA   48
                                      NEW YORK CITY   49
                                      TOKYO   50
                                      TAIPEI   52
                                      LAGOS   55
                                      GLASGOW   54
                                      MOSCOW   56
                                      REYKJAVÍK   57
                                      MONTREAL   58
                                      BAGHDAD   59
                                      SEOUL   60

                                      NOTES   62
                                      ACKNOWLEDGEMENTS   63
                                      IMPRINT   63

FOREWORD

DOPPELGÄNGER AND THE GLIMPSE
DIGITAL TWINS, URBAN STORIES, AND DRAWING CITIES

The statistician George Box famously said "All models are wrong, but some are useful".[1] Box's adage captures the maddening pull of modelling and mapping and visualising: a beguiling sense that we can capture the complexity of the world around us and reduce it down to some essential meaning that can be understood, calibrated, controlled.

As a result, city managers are obsessing over the idea of the "Digital Twin"[2]: a simulation of a city, constructed from data, real-time and otherwise. The Twin seems possible due to the now-usual bundle of data sources, Internet-of-Things technology, machine intelligence, cloud connectivity, and data visualisation. Theoretically, it is created by data from our cities' streets, and their own heap of disparate components: air quality, scooters, retail spend, spoken language, autonomous vehicles, parking spaces, whatever.

The danger of the model, however, is that it promises too much. The word "data" means "that which is given"—from the Latin *dare,* "to give". This definition suggests that data is *not* the self-evident truth of an object. Data is not fact. It is something we are *given* to work with, something we have to interpret, more akin to a complex process than an entry in a field. As Timothy Morton notes in *Being Ecological,*

> "In order to have a fact, you need two things: data, and an *interpretation* of that data [...] Common talk imagines facts to be things like barcodes that you can read off a thing: they are self-evident. But a scientific fact isn't self-evident. That's precisely why you have to do an experiment, collect data and interpret that data."[3]

An emphasis on interpretation is useful. It shows us that data is essentially more subjective than the word is usually taken to mean. This is not a weakness. Rather, it is its strength. This notion of data as subjective emphasises the work we must do with it. It demands that we seek to examine contexts, to seek out what is hidden as well as obvious, to compare types of knowledge. Data provides a glimpse of reality, but not the full picture.

Yet the idea of "the glimpse" is a useful token for unlocking this practice of understanding. The glimpse is the start of something, the start of the path to understanding. The glimpse alludes to what the unfolding picture may bring but does not, or cannot, tell it all at once.

John le Carré said the only way to write about a place was after visiting it for a day, or after a long life once you'd moved there. This would suggest that the period between the day and the lifetime was somewhat useless, in terms of understanding. It didn't lend more clarity, it just accumulated detail. But how else do we get from the insight of a day and of a lifetime? Le Carré's first day in the city is another glimpse. This glimpse can actually be profoundly insightful about a place. However, the glimpse must be interrogated if we are to go any deeper. It is this spirit we should bring to mapping the city. Data provides a flash of understanding, but we must work with it to produce greater insight.

There is something else in the Digital Twin rhetoric that should give us pause: that other kind of twin, the *doppelgänger*, staple of many a horror movie.

This German word is relatively recent coinage; the concept itself has a longer history. It wasn't always such an ominous concept. The version in Finnish mythology, the *etiäinen*, is a spirit that all places, things, and people have. This image or impression goes ahead of a person, doing things the person in question does later, essentially a kind of *déjà vu* in reverse. It lends a peculiar feeling that something is about to happen, and yes, in a typically Finnish mode perhaps, that a bad year is about to come. But not necessarily. As an apparition, it is a kind of working experiential model of the future. This reading of *doppelgänger* echoes the idea of the *glimpse* as far-from-perfect knowledge. It provides a hint of what could be around the corner, depending on what we do next, how we grapple with the complexity of taking action.

The architectural historian Robin Evans wrote that drawing lies "along the main thoroughfare between ideas and things".[4] The kind of abstract modes of visualisation seen in this book usefully hover in-between Morton's idea that there is data, and interpretation of data. They provide a glimpse, but imply work to be done. The tiles in *Cities Squared,* with their flat projection loosely mapped, avoid isometric projection's dalliance with pretend spatial composition, with faux-urban design. They do not suggest that they can be copy-pasted to comprise a community, to solve the problem of a city. But they might tell us where to show up, making digital twins legible via glimpses, providing hooks for multiple conversations and interactions, and dropping data into the everyday-complex context of the street.

The glimpse, then, can be a very useful resource. *Cities Squared* provide such glimpses. The "rational units" of city in this booklet are artfully visualised at just the right level of conceptual altitude to be recognisable, borrowing the saucy wiggle of the River Thames burned into our brains from 6,000 episodes of *EastEnders,* yet also clearly unrepresentative. Each is a glimpse, and so they are hugely promising as a set of tokens to discuss and develop, comprising a generously open visual toolkit for telling stories about cities.

— Dan Hill
 Director of Strategic Design at Vinnova, Sweden

---

1 G. E. P. Box, "Robustness in the strategy of scientific model building", in *Robustness in Statistics,* edited by R.L. Launer and G.N. Wilkinson, Academic Press, 1979.
2 https://en.wikipedia.org/wiki/Digital_twin
3 Timothy Morton, *Being Ecological,* Pelican Books, 2018.
4 Robin Evans, *Translations from Drawing to Building and Other Essays,* Architectural Association, 1996.

INTRODUCTION

Data provides solutions to the most pressing challenges that cities face in the 21st century. Advances in data technologies mean that city managers can capture rich data associated with transport, pollution, and social mobility. But how can we visualise this data, and tell stories with it? How can we use maps to tell these stories?

There have been advances in digital display technology for geospatial data, making it much easier to plot urban data on maps. However, most city maps still remain faithful to real-world geographic scale (or at least given the impression of doing so—all maps transform the world they represent in some way). The issue with geographical fidelity is that lay map-readers tend to conflate size with importance. For example, the public tended to misread maps associated with US elections and the Scottish independence referendum, where large but sparsely populated voting districts seem more important than smaller, but more populous districts. We sought to solve this problem by using "equal-area cartograms"— maps in which all districts are made the same size, meaning each district has the same visual weight (we will refer to these cartograms hereinafter as "square maps"). Square maps are usually only used to represent countries or continents—no one makes square maps of cities. This is partly because there wasn't always so much data on cities, but also because reporting tends to look at larger, national or supra-national patterns. Today there is a lot more granular city-level data available. We had a hunch we could bring the square map down to the scale of the city, and revolutionise how we visualise city data.

Scottish independence referendum results showing counties in which more than half the population voted to leave the UK (shown in blue). The map suggests that only a tiny fraction of the Scottish population voted to leave however the total counts were 2M remain and 1.6M leave.

ORIGINS

This project was born in London, where we were trying to find ways to help the public engage with the huge amounts of data being collected on their city. Existing ways of accessing or representing this data were woefully obtuse. These experiments produced the original London Squared map. This first square city map was a static illustration that demonstrated a concept, rather than functioning fully as a live tool. As the project developed, we played with this new kind of map, exploring what else it could do. We created a simple web app for generating a data-based choropleth.[1] We created a photo collage composed of images pulled from Instagram based on

The original London Squared map that was commissioned by the Future Cities Catapult (Dan Hill, who wrote the foreword, was the director at the time).

9

INTRODUCTION

location tagging. Later, we updated the web tool to make it easier for other people to create their own versions.[2] During this period of play, we also began to apply the concept to other cities, to see what would happen. This book is the result of those experiments.

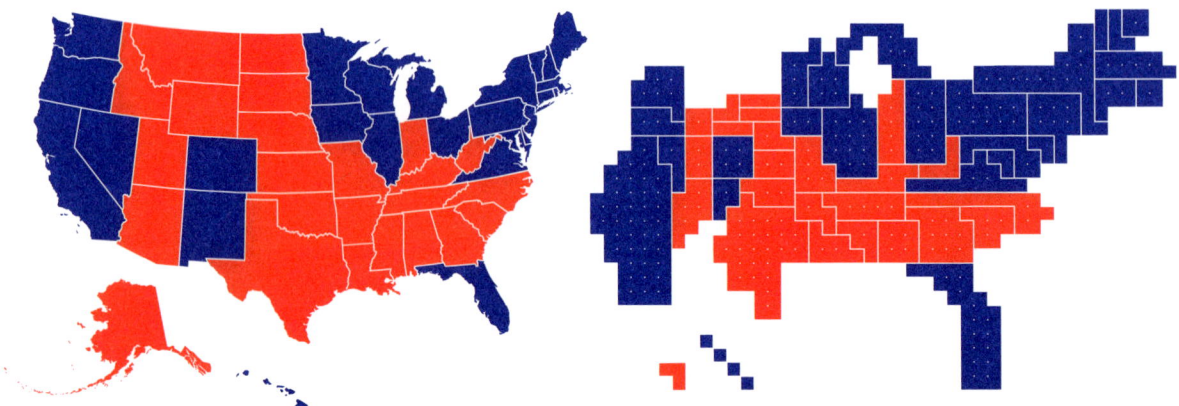

Geographic US election maps are problematic because they can give the false impression that a vast majority of the country voted Republican when in reality the totals are nearly 50/50.

A semi-geographic map that resizes and redraws state shapes to represent each state's number of electors may be unfamiliar but the impression it creates is less misleading in the context of the presidential election.

"RATIONAL" UNITS

We wanted to create maps that would allow district-to-district comparison. We knew that differences in geographic size of districts tends to distract map-readers. To neutralise this problem, we chose to standardise the geographic area of each city unit (by this we refer to boroughs, wards, administrative divisions, etc.—we'll use the generic "district" hereinafter). In the maps in this book, city units are represented by equally-sized "cells." The choice to standardise the size of different districts really is the heart of this project.

When we make the size of districts consistent, the reader can focus more clearly on the relevant data we are seeking to communicate (like, for example, population density). They will no longer be influenced by geographic scale. Equally sized map cells also provide a useful canvas for displaying other forms of data visualisation, such as bar charts, line charts, or pie charts. Such charts sit awkwardly on traditional maps. Clearly, moving away from real world geography opens up a host of representational opportunities. It helps us to use maps to tell a wider variety of stories.

10

SHAPING PERCEPTION

Standardising the size of each city district does present challenges. Square maps can distort the physical shape of a city and the relative geographical location of districts within that city. Sometimes these distortions are too extreme—at some point, a square map can just stop looking enough like the city it is meant to represent. Sometimes districts seem to join that shouldn't, and sometimes districts are far apart when inhabitants know full well they are adjacent. Sometimes the square map annihilates the iconic outline of a city (how would we recognise London without the famous bends of the Thames?)

While our square maps distort physical geography, so does every map projection. No mapping system is perfect. The most commonly used map projection, Mercator, became popular because it represented any path with a constant bearing as a straight line. This was a boon to sailors in the 16th century, but it also skews geographic area heavily toward the northern hemisphere, creating the perception that the North America and Europe are larger than they actually are, relative to South America, Africa, and Australia.

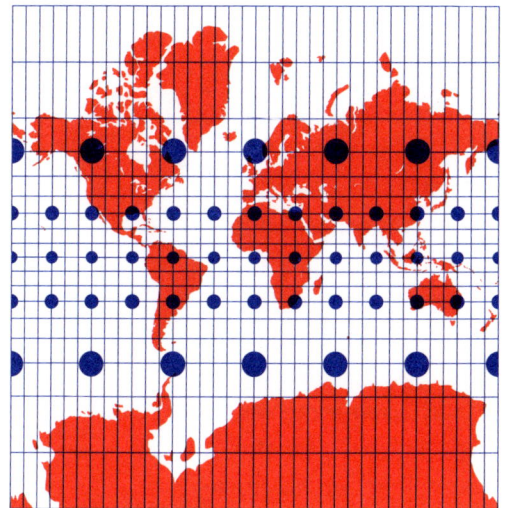

Mercator projection maintains the shapes of landmasses but distorts their sizes. Size inflation increases as you move away from the equator. One result of this inflation is that Greenland and Africa appear to be roughly the same size, when in reality Greenland is about 14 times smaller than Africa.

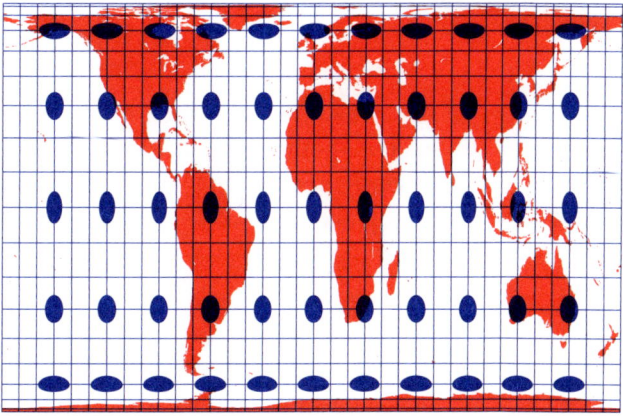

Gall-Peters projection maintains the correct relative size of all landmasses but distorts their shapes.

The dots on the maps above are called Tissot Indicatrices and they represent the amount and kind of distortion at any point on a map. For instance, in the map seen above the dots all have the same area but are stretched vertically or horizontally to different degrees depending on where they fall on the map, showing how the corresponding map locations have been distorted.

INTRODUCTION

During the Cold War the USA used the inflated size of the USSR on Mercator projection maps to maintain a high level of national anxiety about the scale of the threat posed to America.

In Mercator projection the USSR was approximately five times the size of the USA.

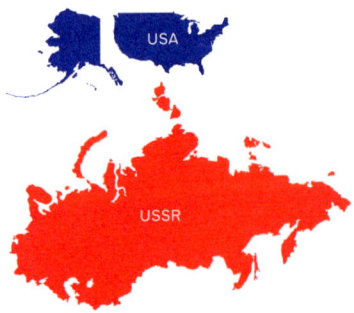

In reality the USSR was closer to 2.25 times the size of the USA (22.4M km² versus 9.83M km², respectively).

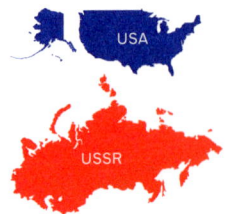

Cartographers make choices about trade-offs like this every day in their work. We have made ours. As Mark Monmonier points out in *How to Lie with Maps*,

> "A good map tells a multitude of little white lies; it suppresses truth to help the user see what needs to be seen. Reality is a three-dimensional, rich in detail, and far too factual to allow a complete yet uncluttered two-dimensional scale model. Indeed, a map that did not generalise would be useless. But the value of a map depends on how well its generalised geometry and generalised content reflect a chosen aspect of reality."[3]

Before GPS was widely available and everyone had a supercomputer in their pocket, maps were made by hand on paper, and the creation of a new one was a major endeavour. Due in part to the amount of work involved, governments, corporations, or cartographers always brought a clear agenda to any map project. Modern technology makes map making is less arduous. It has lowered the stakes, and enabled us to play.

RECOGNITION

We think it is important that residents can recognise their city in our maps. We did learn various ways we can modify the square map to alleviate distortions. We learned to work creatively with the arrangement of districts to align them appropriately. We also learned the value of creating "visual hooks" that evoke the city's outline. A visual hook is a deviation from the standard square grid, based on some aspect of topography (a river, a bay, a coastline) that alters one or more map cells with the intention of making the map a bit more recognisable to the viewer. However, this "visual hook" sometimes eludes us. Acceptable square maps cannot be made for all cities (we conceded defeat with Mexico City and Barcelona, for example).

What counts as acceptable is difficult to define. We often say a map "feels right" or "feels wrong". We say this because, beyond all the technical work, intuition is important. When designing a map like our square maps (which is to say, when working with obvious and extreme abstraction) it is important to keep in mind what these maps are for. They are simplifications that need to evoke a city at a glance. As such, a viewer needs to be able to at least guess the city's identity with a fair degree of certainty. If the square map of a city is a total mystery, it is a failure, and we reject it.

Though we take our cue from established districts, it is worth noting that these divisions are not necessarily rational, good, or fair.[4] Dividing urban space into administrative regions is often arbitrary—there is rarely anything intrinsic about how cities are parcelled out into administrative subregions. And each city does it differently. District boundaries might follow the topography of the city (roads, canals, rivers, etc.), or they might be overlaid as a geometric grid. The stability of district borders ranges from rigidly fixed (New York City's districts have not changed since 1898) to wildly erratic (Montreal has changed its district layout three times since 2000). Additionally, what is considered part of a city can range from expansive ("Auckland" covers some 5000 km²) to restrictive (Copenhagen is only 88 km²). Despite this variation, each city does collect data at the level of these districts, and these are the units we must work with. And despite the variation, these areas are all typically small enough to be able to show how patterns change across the landscape of the city and large enough to be read quickly without a magnifying glass.

A geographic map of Stockholm versus its square map counterpart. The square map remains recognisable even though there are a number of small inaccuracies. The overall shape of the square map is similar enough to actual geography and the insertion of the diamond-shaped visual hook helps identify the city centre, resulting in a useful abstraction for plotting many kinds of data sets.

Strategies for how to divide a city into administrative units are as varied as the cities themselves. Some cities rely primarily on natural topographic features to define the edges of districts (e.g. New York City). Other cities have districts based primarily on historical neighbourhood boundaries (e.g. Toronto).

New York City, USA
5 districts
784 km²

Havana, Cuba
15 districts
782 km²

Toronto, Canada
25 districts
630 km²

Santiago, Chile
35 districts
641 km²

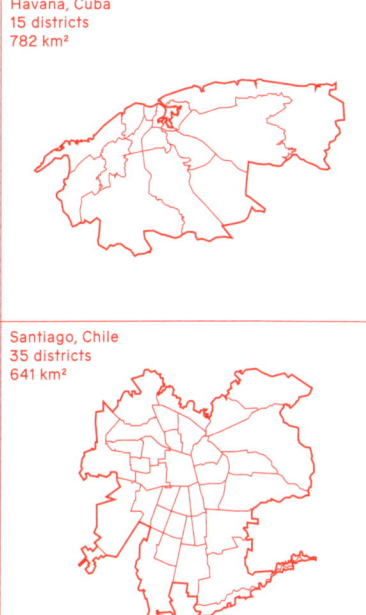

The maps to the right all have a scale of 1:1,111,111

13

INTRODUCTION

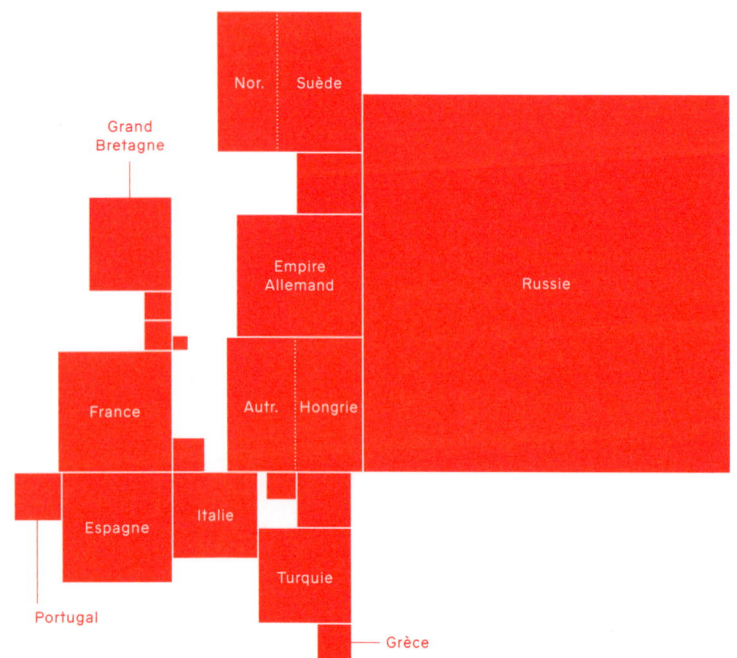

Cartograms arranged by hand, such as the one above by Émile Levasseur from 1868, can be tailored to the data available to make recognition easier but the process can be time consuming for large numbers of areas.

Dorling cartograms (1996) take a more abstract approach allowing for clear comparison between areas but the relationship to position, shape, and topology of areas can become tenuous. It's clear Glasgow is bigger than Edinburgh but without the labels you wouldn't guess that this is a map of Scotland.

ON CARTOGRAMS

A cartogram is a kind of map that changes the size, and sometimes shape, of geographic areas (e.g. county, province, borough) based on some measured dimension such as population or GDP. In the *Cities Squared* project, we worked with administrative districts.

Cartograms can work in a number of ways. Some cartograms convert all map units into the same shape (our square maps work this way), whilst some retain the original shape but scale them up or down. Some retain the original shape of the units and scale them, but leave gaps in order to conjure the city's outline. And some create crazy liquid-looking blob shapes that look like marble endpaper. None are perfect, and each option involves a trade-off. Some have a clearer relationship to standard geographic representation, some are easier to read as a set of quantities. The cartographer must decide which kind of map is best suited to telling their story.

The earliest cartograms are often credited to Émile Levasseur, who used them to illustrate his economic geography text book of 1868. Levasseur arranged rectangles to form a morphologically recognisable map of Europe, and scaled and coloured these rectangles according to various schemes. The early 20th century saw a number of attempts at cartograms which retain the characteristic shapes of administrative areas whilst distorting their size, but it wasn't until the adoption of computers in the latter half of that century that more complex cartograms became possible; notably through the work of Waldo Tobler, Judy M. Olsen, Danny Dorling, and most recently Gastner and Newman who developed "diffusion cartograms" (which, we feel, look cool but are not actually that useful).

The earliest computer generated cartograms (Waldo Tobler, 1961) retain the relationship between areas and some degree of recognisability at the expense of the ease with which different areas can be compared. The example below dates from 1986.

Gastner-Newman diffusion cartograms (2004), whilst visually arresting, often makes it very difficult to read values or make comparisons. For instance, in the example below, does Texas or New York have a greater size?

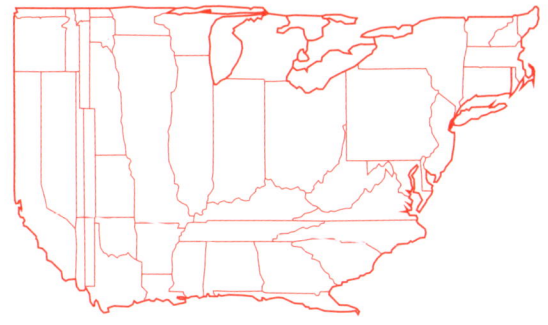

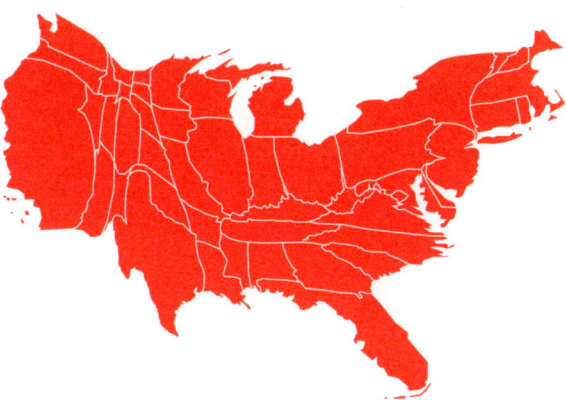

Non-contiguous cartograms make for easy comparisons and improve area recognisability at the expense of showing shared borders.

Recently the popularity of cartograms of all types has blossomed, thanks in part to the efforts of graphics teams at news organisations putting more novel visualisation methods before the public, and in part to the ready availability of production technologies such as ggplot, D3js, and commercial offerings such as Tableau. You no longer need to be a computer scientist or a seasoned cartographer to produce high quality cartograms.

INTRODUCTION

THE DATA-DRIVEN CITY

Cities have long been a source of and place to use data, especially since the industrial revolution. Today many cities produce an astonishing amount of data in the course of daily operations. But numbers are not knowledge. Interpretation provides the leap from raw data to genuine insight. Sometimes data is under-interrogated—we have seen many city officials confuse large amounts of data with actual understanding of what is going on. Data can be made to serve various political agendas, since most datasets can be manipulated to tell a wide variety of stories depending on how they are organised and edited. We, too, tell stories with data in the maps in this book—though, we hope, in better faith. We aim to use maps to help people better understand quantitative data and statistics. But we are always mindful of how easy it is to abuse these tools to grossly misrepresent "the truth". If data is an increasingly important way to "see" our environment, how do you display it in a way that is fair? What does that actually mean? Is that even possible?

We believe alternative forms of cartographic and data display help people make decisions better and faster. The strength of the square maps in this book lies in their simplification and abstraction of data, suggesting new perspectives and understanding for those who manage cities and city life. They are not the only way to show local data, but a new kind of tool for understanding what data is telling us about the world.

1   A choropleth is a map that uses some colour/shade or pattern to visualise some dimension (e.g. population size, birth rate, number of bakeries) on a zone-by-zone basis—by definition you can't make a choropleth without cutting a map up into areas. Typically these areas are some official unit (e.g. a state, province, voting ward).
2   See: https://tools.aftertheflood.com/londonsquared
3   Mark Monmonier, *How to Lie with Maps* (second edition), University of Chicago Press, 1991/1996, p.25.
4   Where district lines fall is decided by politicians who often have selfish goals. The drawing of voting districts in the US is a great example of how the process of creating administrative regions can be twisted toward political ends.

PROCESS

PATTERNS

We began by examining the various ways city districts are arranged, seeking to identify common patterns. This research exercise helped us understand which urban forms were best suited to this project.

Radial
The radial structure is probably the most common city form and consists of an urban core surrounded on all or most sides by larger districts. District size typically increases as one moves from centre to periphery. Milan is the purest example of this kind of radial city.

In a common variation, the shape is asymmetrical. In these radial cities, there is a definite city centre from which peripheral districts radiate, but most of the growth happens in a single direction. Mexico City is a pretty classic example of this structure. Typically this kind of asymmetric growth is influenced by topographic features, such as spreading toward a river or being hemmed in by a mountain.

When translating these cities into square maps, the outer districts sometimes don't connect, when they do in reality. This tends to happen when the districts vary in size a lot. It means that districts get shifted into odd positions, and this effect becomes more and more extreme towards the periphery of the city.

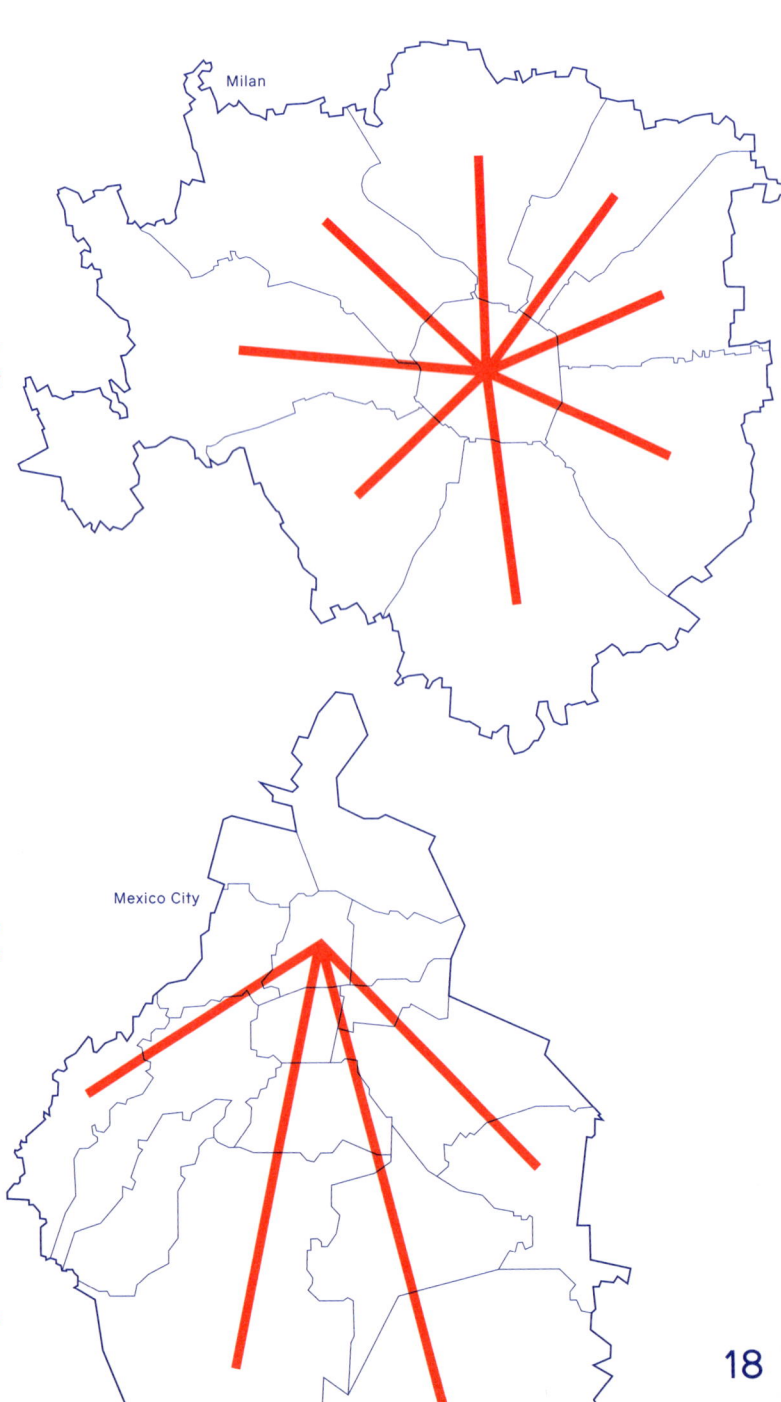

18

St. Petersburg

Arc
Often the result of a city spreading along or around a topographic feature, this arrangement is very common for cities that sit on a concave bay, along a bending river, or in a valley. Many cities of this type also feature a protrusion that deviates from the primary sweep of the city. This city shape is often a good candidate for a square map because the arc will serve to provide a natural visual anchor. In square maps of these cities the outer districts can get separated from each other, but this is not always extreme enough to discard the map.

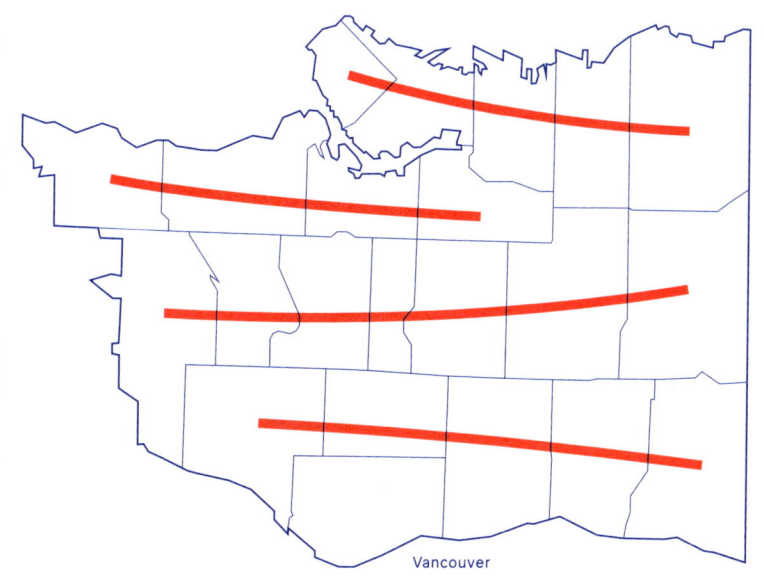

Vancouver

Linear
Linear structures come closest to a pre-made grid structure, though most instances of this structure show signs of physical topography or politics in their district arrangement. This structure is most common in North America (e.g. Toronto, Chicago, or Denver) where cities don't have a medieval core or long history of accretion to contend with. This city shape can be surprisingly resistant, and it is interesting to note that Vancouver's very slightly irregular grid structure actually makes it very hard to create an effective squared map for it.

SHAPES

After we had a better understanding of the way districts were arranged, we began exploring what shapes were best suited to representing the individual districts themselves. All of the shapes we tested were polygons. The core idea of the project—equalising the size of each district so that data can be visualised without geographic distortions—meant that all of the shapes on any given map be the same (or nearly so, allowing for small alterations to create a recognisable landmark). We considered allowing each map to be built from whatever shape worked best for the city in question, but this would make the maps less comparable. After we chose a few test cities to work with we went about exploring what shape would be most useful for expanding the project. We tested triangles, squares, pentagons, hexagons, and octagons. In the end we returned to the square as the best and most useful shape because of its visual simplicity and usefulness with data display, but all the other shapes offered advantages.

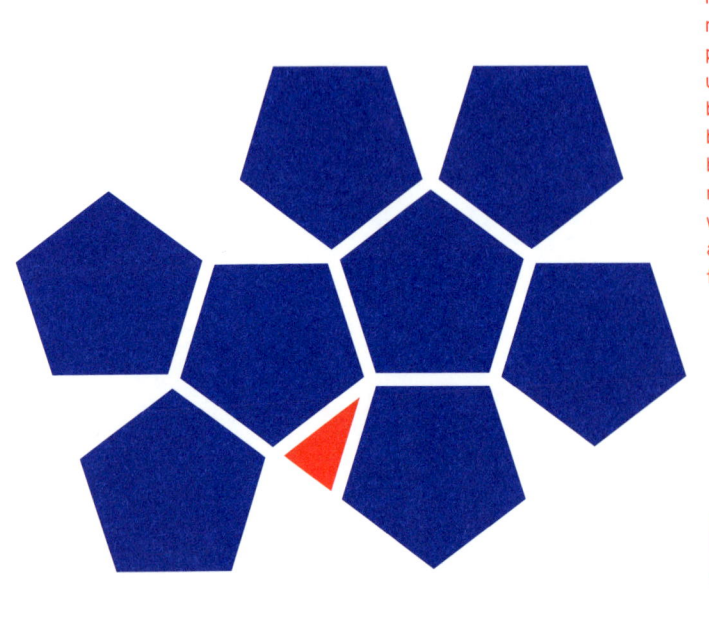

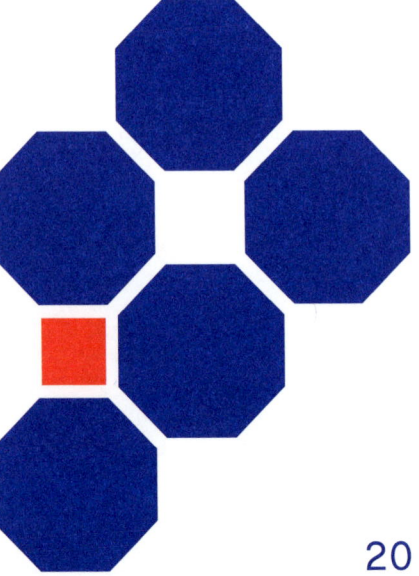

Pentagons and octagons both allow the map maker to create beautiful geometric patterns. If we had pursued the path of using a variety of shapes, choosing the best for each individual map as dictated by the city in question, they may have been viable options. Unfortunately, neither can be arranged into an even grid without gaps, and so we ruled both out, as they would prevent us from creating tightly gridded maps.

20

Diamond

Diamonds can be set into a tight grid, but this grid can be arranged in a variety of ways, including diagonally. This is useful when creating a square map based on irregular city districts that do not line up nicely. The diagonal grid allows greater flexibility, and avoids the distortions that we generate when we force a less than rectilinear real world into uniform units. Unfortunately the diamond is very hard to work with in terms of data overlay, so in the end they were ruled out.

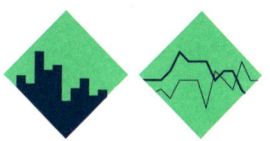

Hexagon

Whilst the square triumphed, the hexagon was the worthy runner-up. Hexagons lock up nicely into a regular grid, and are very flexible in how you arrange them. Unfortunately they are far inferior to squares for overlaying data (they can only really accommodate pie charts, heatmaps, or dot patterns).

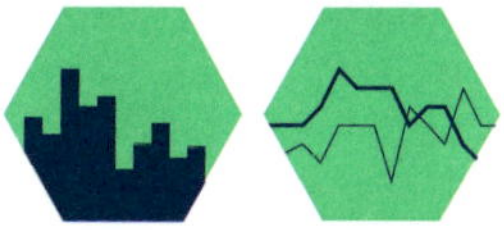

Square

Squares were the default option after the London Squared map. We still experimented with a variety of shapes to make sure it was the correct choice. In the end squares proved the most flexible for map making (despite the various trade-offs involved with forcing city spaces into a 90° rectilinear grid) and were the best shape for data overlay by a country mile.

 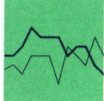

21

LIMITS

The process of translating city maps into squared maps can be tricky. Cities come in all manner of shapes; creating rationalised, square cartograms from these shapes is sometimes straightforward and sometimes extremely convoluted, or just not possible. In the next section of this booklet we have a large set of maps that we think are successful, but the system does not work for all cities, and there are common patterns to be found in how the system fails. In all of the following examples, when trying to arrange the squares, there is typically a nagging feeling that it almost works, if you could just get a certain detail to work out, and yet somehow it never quite comes together. In the maps we think are successful there are often trade-offs but none of them felt like compromises that break the map.

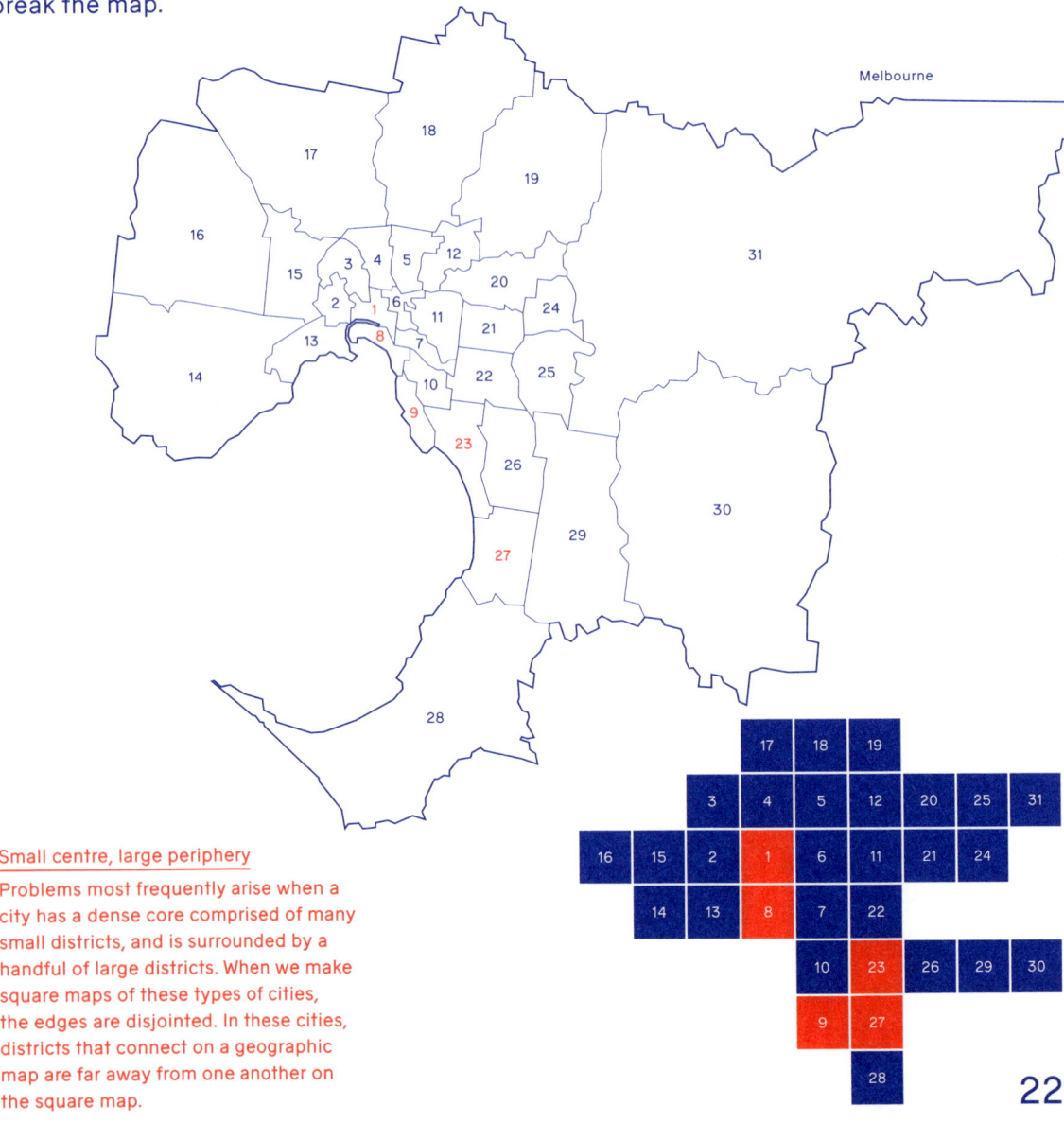

Small centre, large periphery
Problems most frequently arise when a city has a dense core comprised of many small districts, and is surrounded by a handful of large districts. When we make square maps of these types of cities, the edges are disjointed. In these cities, districts that connect on a geographic map are far away from one another on the square map.

The one-to-several problem

Problems arise when two (or more) districts share a border with one larger district. This is especially troublesome when it happens near the centre of the city. In these cases, it becomes difficult to arrange any districts further toward the periphery, often resulting in districts along one of the city's outer edges not touching when they should. This situation is especially prevalent in cities that are gridded and seem highly regular.

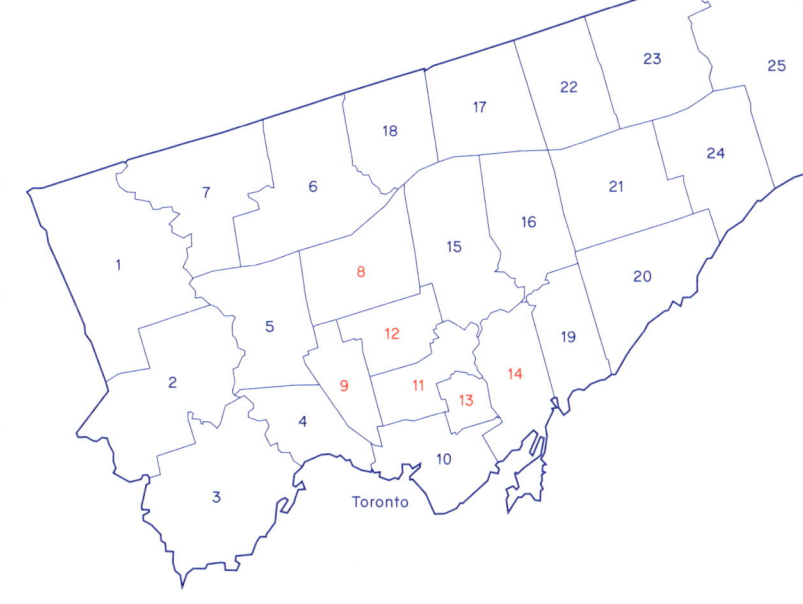

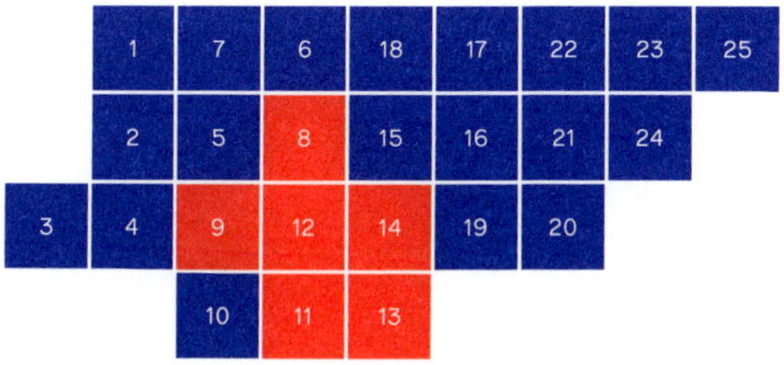

Toronto is interesting because large areas of it can be translated quite effectively into a square map, but the centre of the city has a geographic arrangement that effectively rules it out as a viable candidate—issues such as ward 13 being pushed to the coast when it should be inland. No matter how you move the squares around the result has at least a few districts that should line up being out of place.

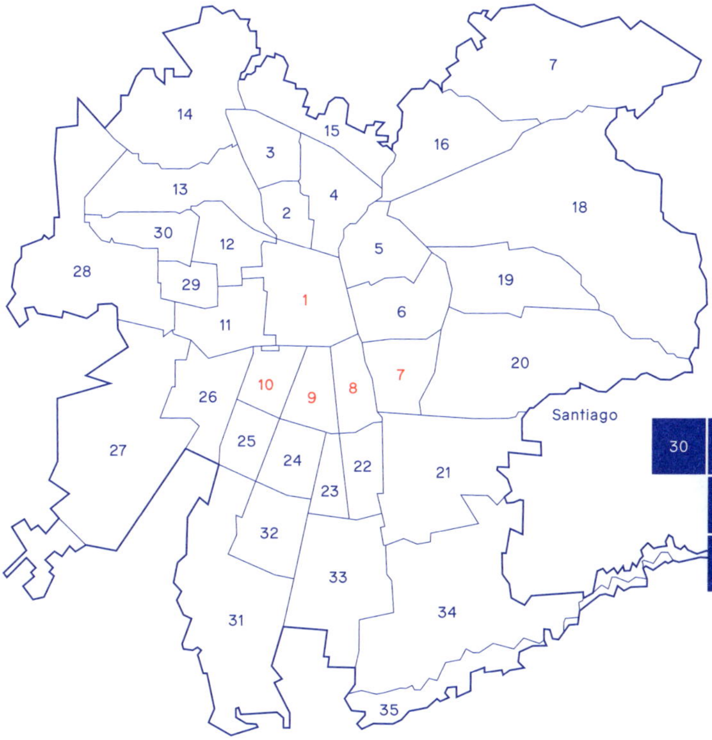

Like Toronto, Santiago *almost* works as a square map, but the way that district 1 borders district 10, 9, and 8 has knock-on effects that are impossible to avoid. The outer edges have significant problems (e.g. district 30 and district 7 being pushed to the outer edges of the map).

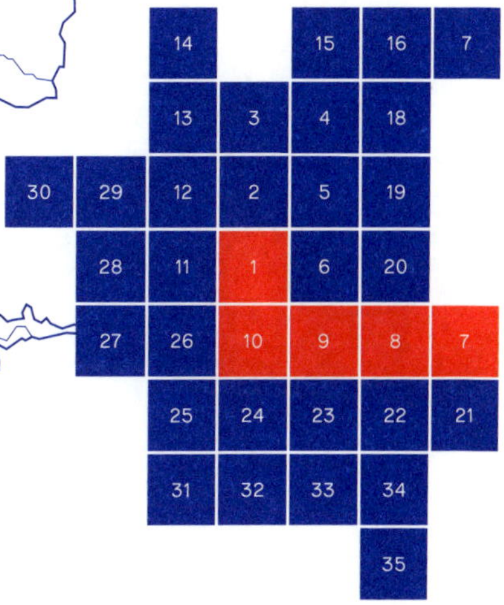

LIMITS

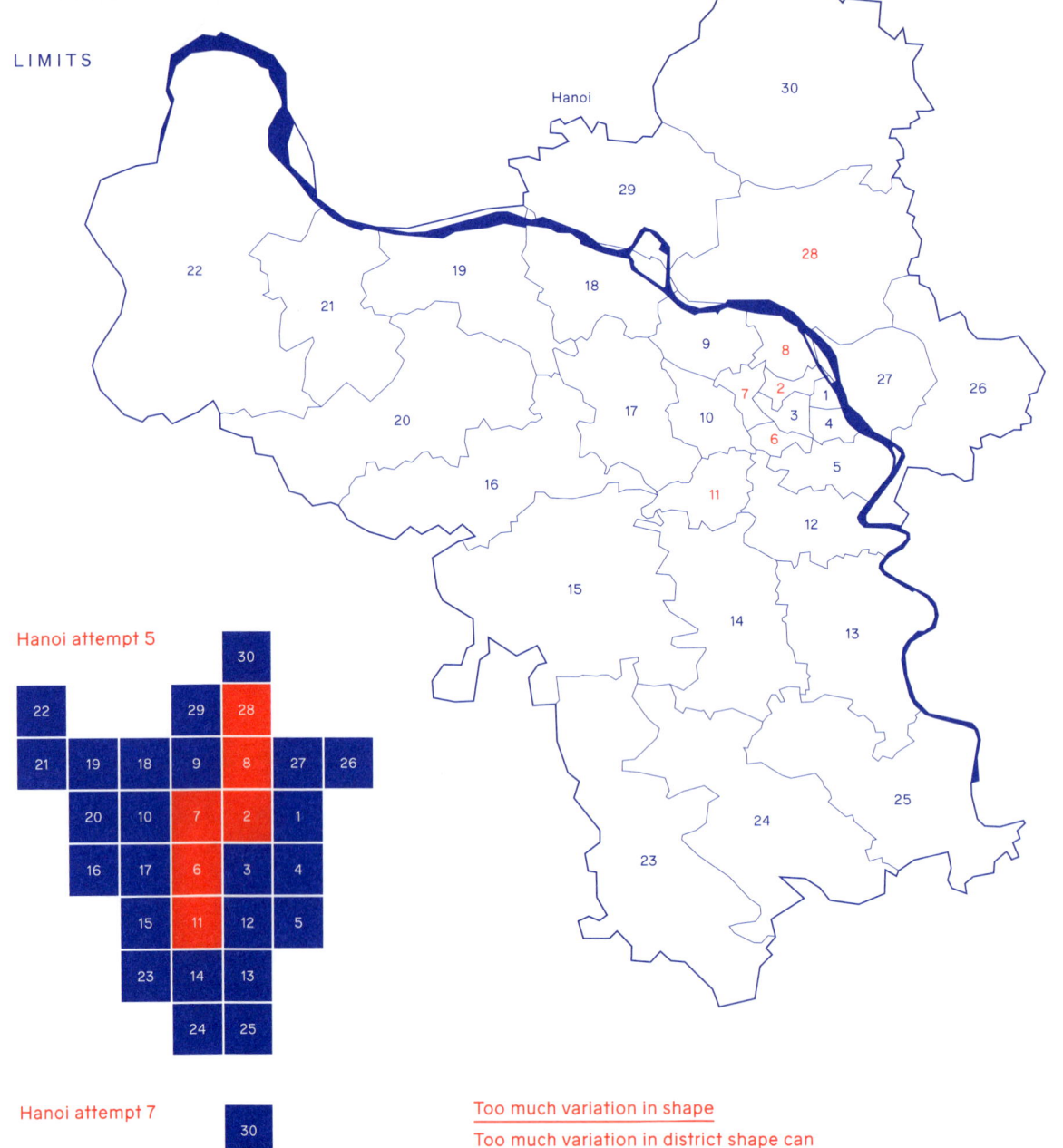

Hanoi attempt 5

Hanoi attempt 7

Too much variation in shape

Too much variation in district shape can also cause unacceptable distortions. Variations in shape make it tricky to line up squares in a way that represents geographic reality. We would have to fiddle, and make compromises. This process sometimes ends in frustration. It can often feel like we are very close to getting the map right, but ultimately we must accept that the compromises are just too great. Hanoi's districts are a chaotic jumble of many different shapes, and we drafted over ten different versions of a square map—none were acceptable (though all were almost acceptable)

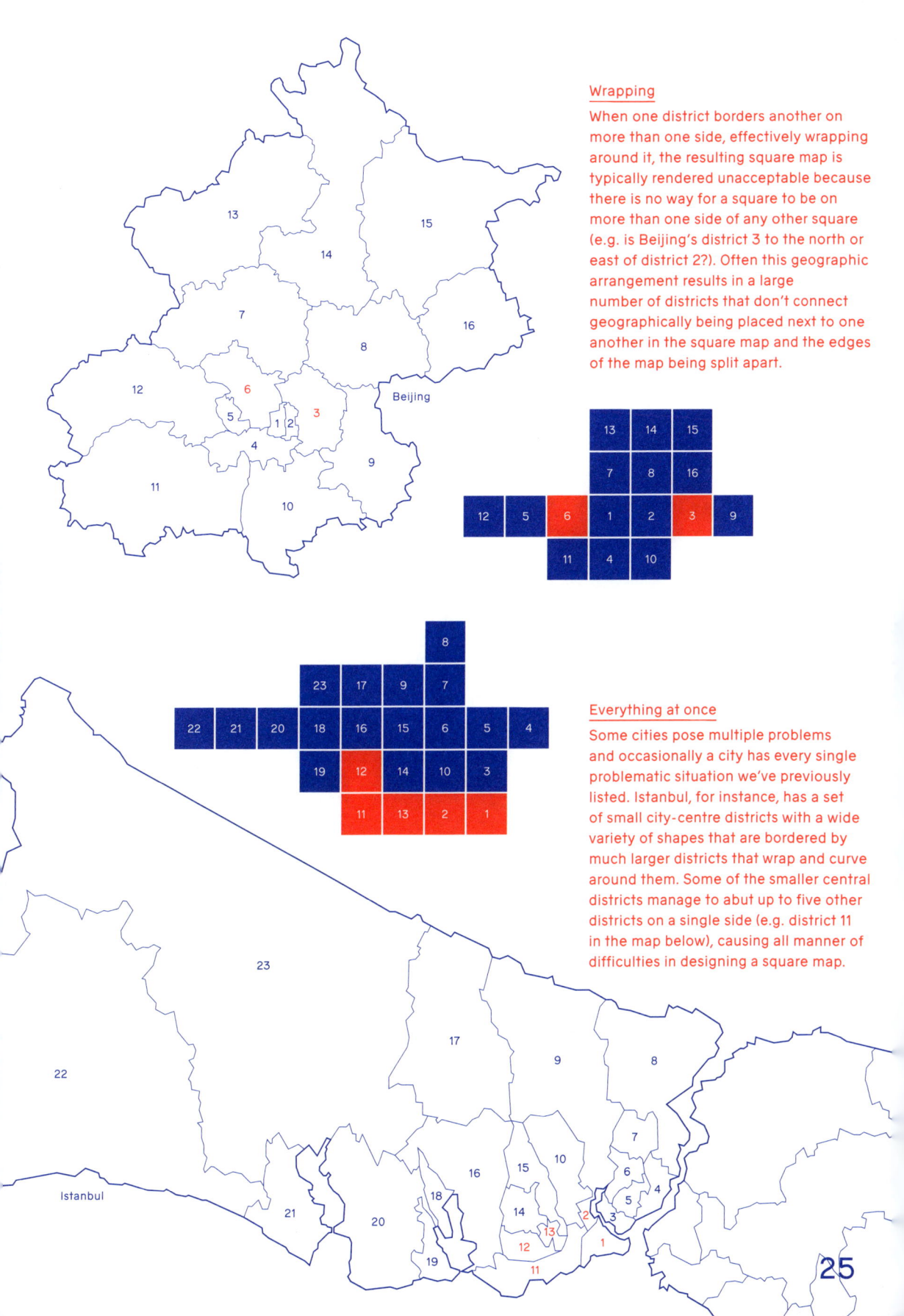

Wrapping

When one district borders another on more than one side, effectively wrapping around it, the resulting square map is typically rendered unacceptable because there is no way for a square to be on more than one side of any other square (e.g. is Beijing's district 3 to the north or east of district 2?). Often this geographic arrangement results in a large number of districts that don't connect geographically being placed next to one another in the square map and the edges of the map being split apart.

Everything at once

Some cities pose multiple problems and occasionally a city has every single problematic situation we've previously listed. Istanbul, for instance, has a set of small city-centre districts with a wide variety of shapes that are bordered by much larger districts that wrap and curve around them. Some of the smaller central districts manage to abut up to five other districts on a single side (e.g. district 11 in the map below), causing all manner of difficulties in designing a square map.

VISUAL HOOKS

The removal of difference via a completely gridded, "rational" map is a double-edged sword: normalising all district shapes into squares removes size bias in data reporting but it also makes it hard to tell maps apart. Visual hooks aim to re-introduce visually identifiable elements to make it easier for the reader to quickly identify a city, and orient themselves within that map. After some exploration we eventually landed on the idea of "edges" as the most effective type of city element to work with (we also tested other types of landmarks such as buildings, but these didn't integrate well with the square grid base layer). We based our notion of edges as a category of urban identification on the work urban planner Kevin Lynch (see: *The Image of the City*). Edges can take the form of rivers, shorelines, mountain ranges, etc. They are often situated between districts or at the periphery of a city, making them ideal for our gridded maps, which by their nature have a lot of edges to play with.

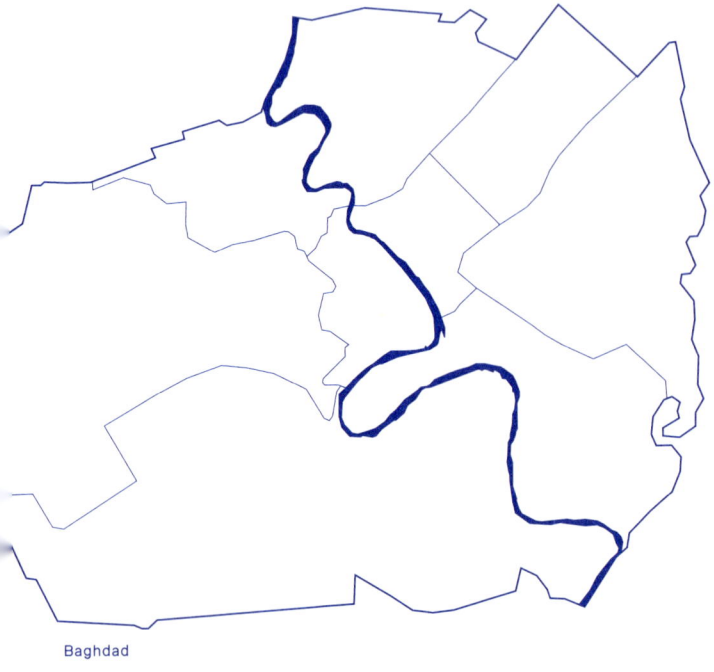

Baghdad

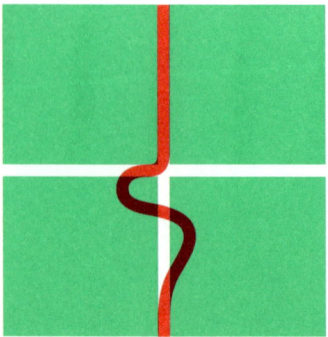

Winding curves

Rivers are strongly evocative. Each is unique. When a city has a river, this river tends to act as a focal point, with districts arranged around and along it. It is important to find a way to simplify a river's form without making it overly generic—this can be a tricky balance to achieve.

26

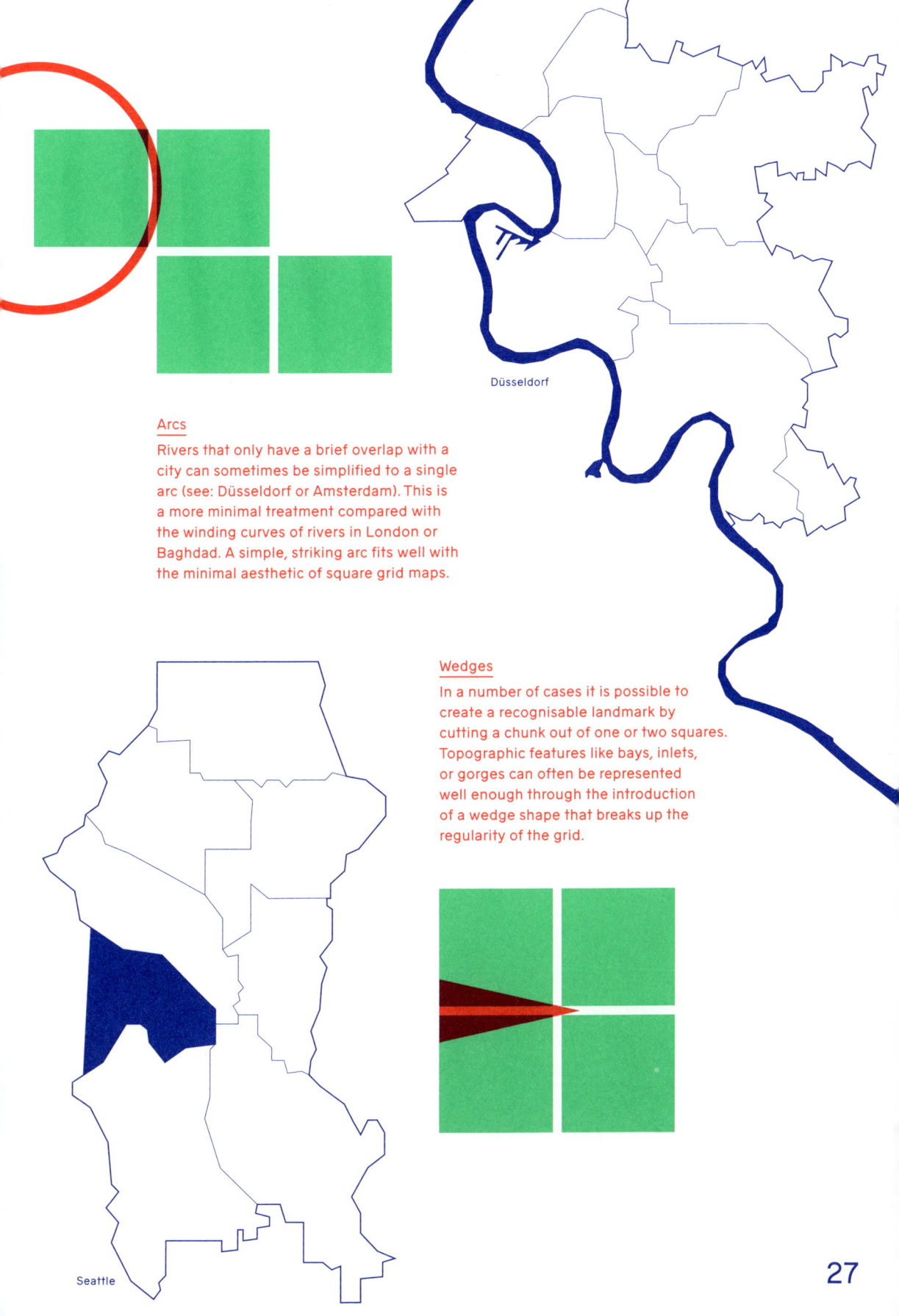

Arcs
Rivers that only have a brief overlap with a city can sometimes be simplified to a single arc (see: Düsseldorf or Amsterdam). This is a more minimal treatment compared with the winding curves of rivers in London or Baghdad. A simple, striking arc fits well with the minimal aesthetic of square grid maps.

Wedges
In a number of cases it is possible to create a recognisable landmark by cutting a chunk out of one or two squares. Topographic features like bays, inlets, or gorges can often be represented well enough through the introduction of a wedge shape that breaks up the regularity of the grid.

VISUAL HOOKS

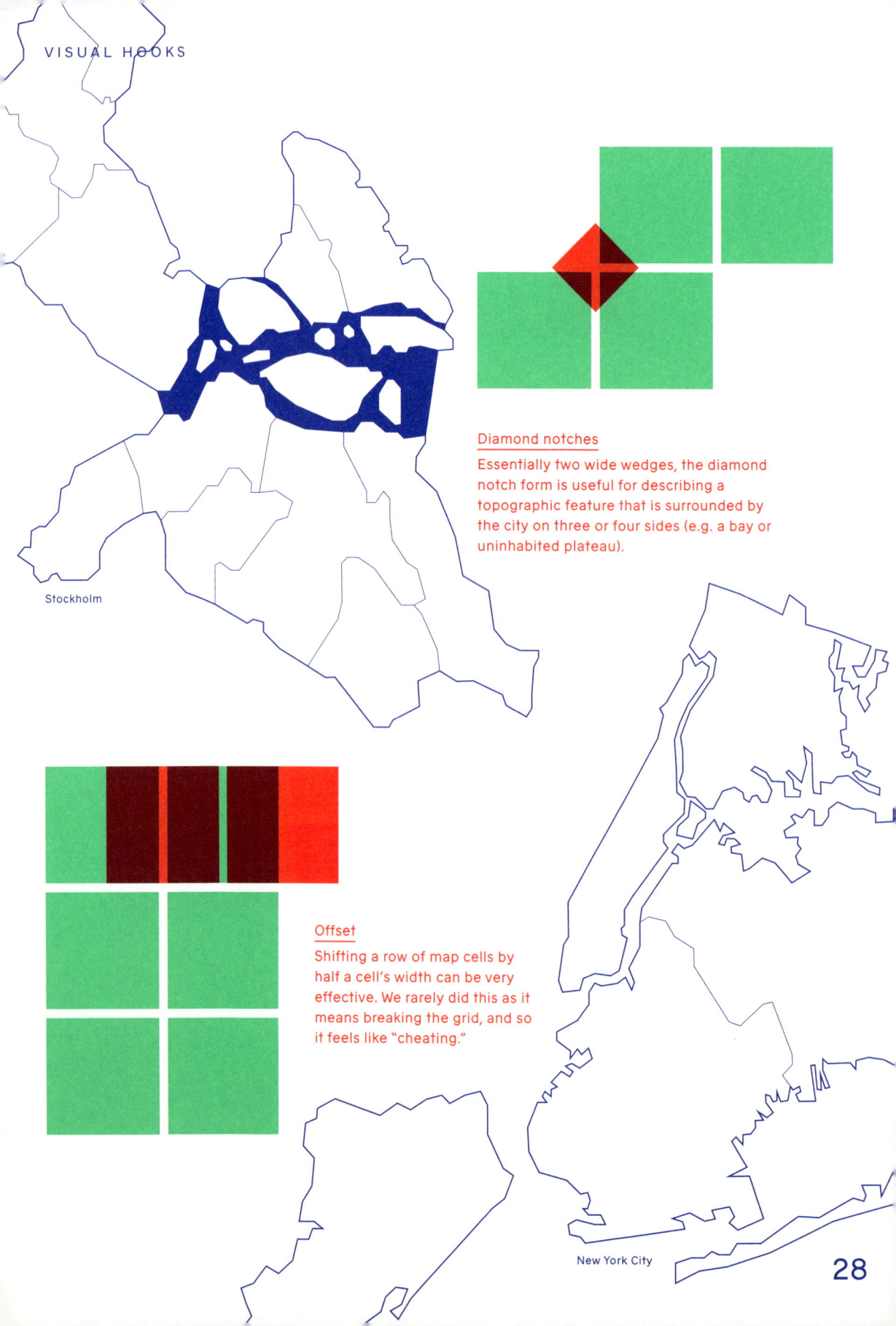

Stockholm

Diamond notches
Essentially two wide wedges, the diamond notch form is useful for describing a topographic feature that is surrounded by the city on three or four sides (e.g. a bay or uninhabited plateau).

Offset
Shifting a row of map cells by half a cell's width can be very effective. We rarely did this as it means breaking the grid, and so it feels like "cheating."

New York City

28

MAPS

4000 km²

The waterline
The overall size of each square map is determined by the number of districts in the city. However, there is no real relationship between number of districts in a city and a city's actual physical size. This means our square maps tend to distort any sense of relative geographic scale. To account for and document this distortion we've included a waterline of sorts, in the form of an overprinted yellow rectangle on each of the map pages, that indicates the geographic size of each city.

LONDON

London marks the starting point of the *Cities Squared* project. Our intimate knowledge of the city helped us arrange the boroughs in a way that made sense and *felt* right. Whilst the precise arrangement of the blocks is not technically accurate, we know we created a map that matched our instinctive understanding of how the boroughs fit together. The river came later, and was crucial in giving the map some personality, and providing a key marker to help readers orient themselves in the map space. The River Thames acted as the very first "visual hook", and the London Squared map taught us that such a device would be crucial for all square maps.

The River Thames is a key visual anchor in the London Squared map. We chose not to include all the river's bends, because this was too disruptive to the square grid. In any case, a simpler shape was actually better suited to evoking the iconic and highly recognisable river. We isolated the area around the Isle of Dogs because of its distinctive double-bend.

The area of London is of 1,572 km², with a population of 8,825,000 people. London is composed of 32 boroughs, each with its own local authority. The local borough councils have operated semi-autonomously since the dismantling of the Greater London Council in 1986, though the Greater London Authority, established in 2000, has restored some central coordination.

ENF	Enfield	CTY	City of London
HRW	Harrow	TOW	Tower Hamlets
BRN	Barnet	NWM	Newham
HGY	Haringey	BAR	Barking and Dagenham
WTH	Waltham Forest		
HDN	Hillingdon	RCH	Richmond upon Thames
ELG	Ealing		
BRT	Brent	WNS	Wandsworth
CMD	Camden	LAM	Lambeth
ISL	Islington	SWR	Southwark
HCK	Hackney	LSH	Lewisham
RDB	Redbrige	GRN	Greenwich
HVG	Havering	BXI	Bexley
HNS	Hounslow	KNG	Kingston upon Thames
HMS	Hammersmith and Fulham		
		MRT	Merton
KNS	Kensington and Chelsea	CRD	Croydon
		BRM	Bromley
WST	Westminster	STN	Sutton

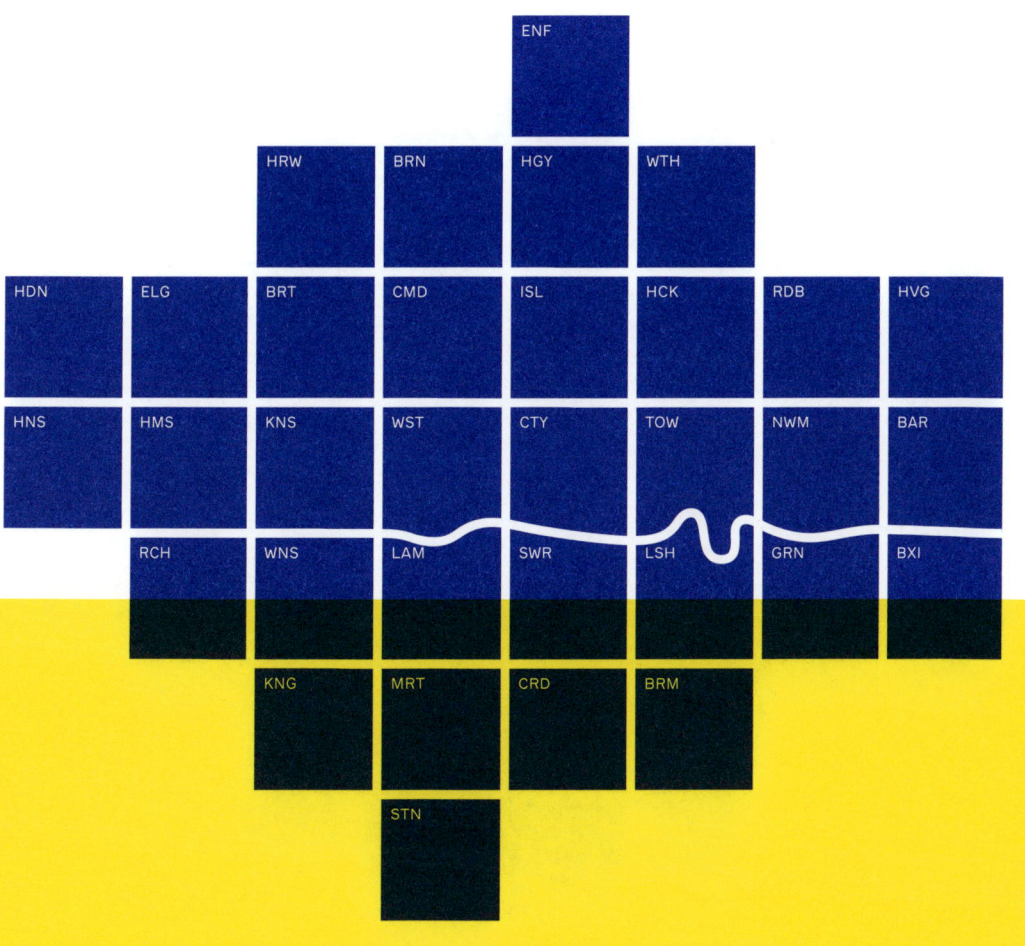

1,572 km²

31

LONDON

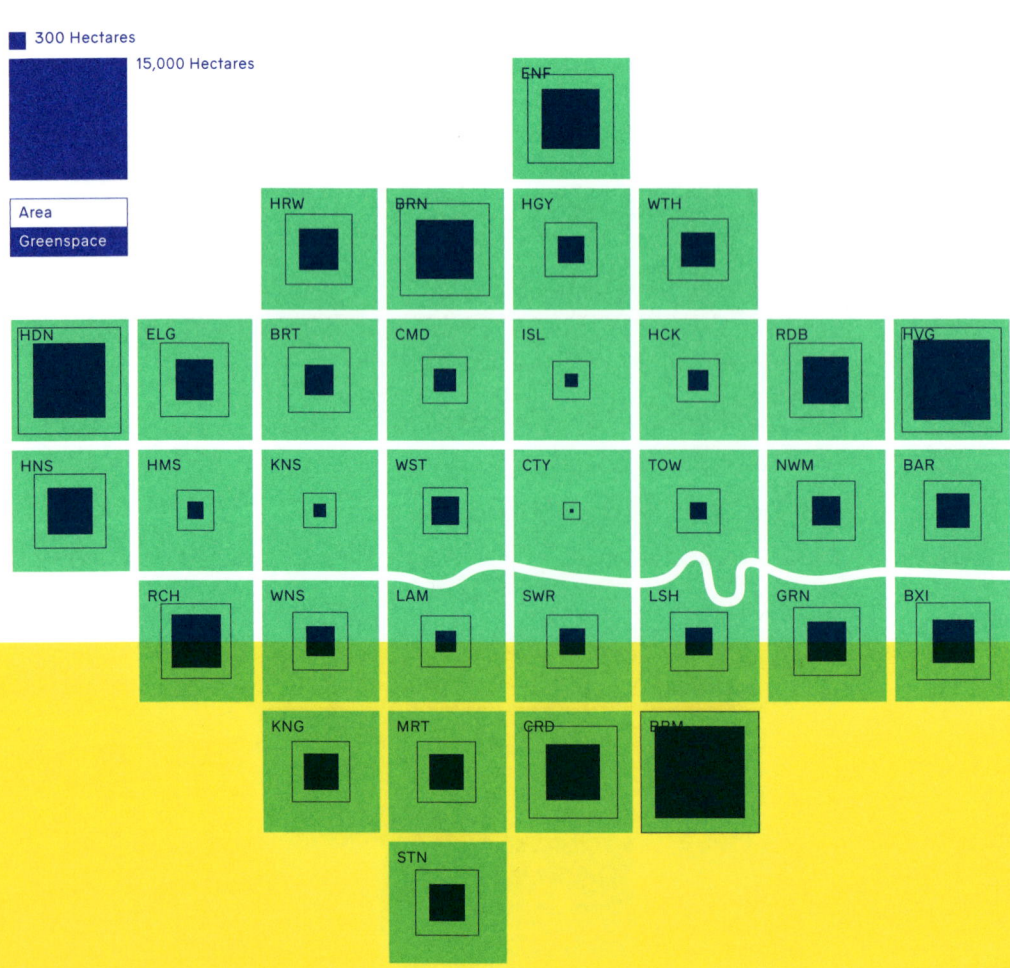

32

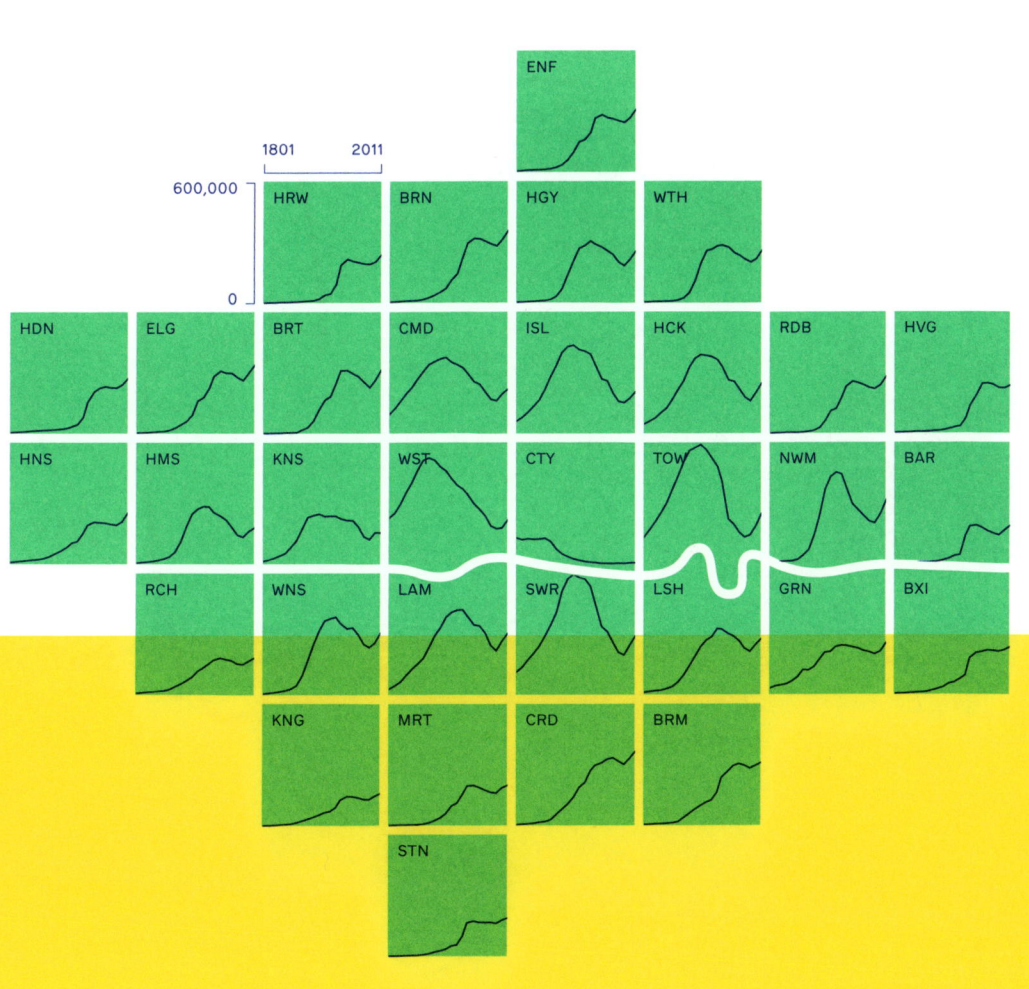

SEATTLE

Seattle is an example of a city that is comprised of a relatively small number of large districts. When we create basic square maps of these kinds of cities, they tend to just look the same as each other. When trying to add a reference point that will give a map individual character, and make the city recognisable, the question is: "how much is enough?"

With Seattle, we learned that the "wedge" form was a simple but highly effective intervention. In the map, we chose to represent Seattle's Elliot Bay with this new wedge shape. This Bay asserts a significant presence in Seattle—it dramatically cuts into the urban space, and abuts Central Seattle. As such, it was an obvious feature to represent in order to make the map more recognisable.

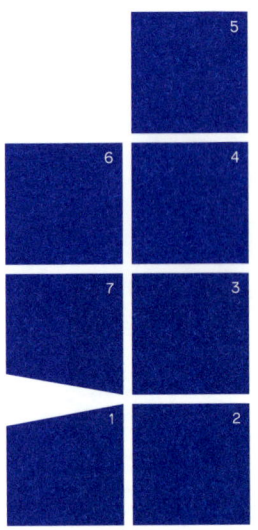

In Seattle, districts are simply named from 1 to 7. This order is pleasing—but it does seem strangely too orderly to those who are unfamiliar to Seattle. Aside from District 5, the northernmost district, Seattle's districts are arranged in grids, more or less. Whilst this often makes our job straightforward in terms of arranging districts, we did move district five slightly to the east to better evoke the overall city shape (but it could also have easily been placed north of district six).

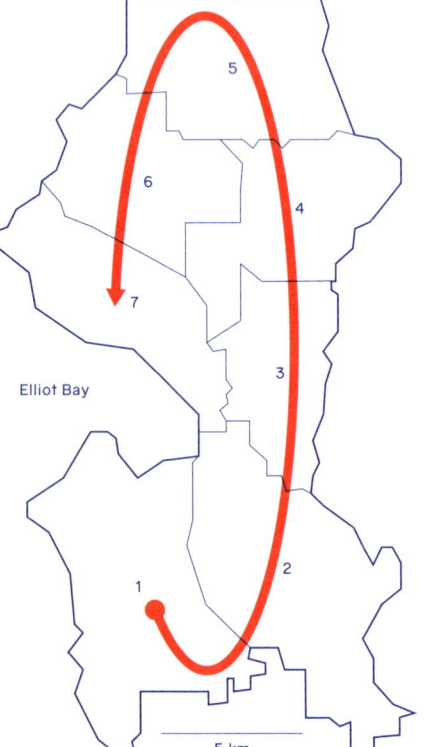

1 Council District 1
2 Council District 2
3 Council District 3
4 Council District 4
5 Council District 5
6 Council District 6
7 Council District 7

The area of Seattle is of 217 km², with a population of 3,867,046 people. The City of Seattle is divided into seven Council Districts, with a Councilmember representing each district.

34 217 km²

COPENHAGEN

A couple of factors make Copenhagen tricky. One factor is the central neighbourhood Frederiksberg. This area is completely surrounded by multiple city districts. But it is not actually part of the city—it leaves a gaping hole in the map. The second factor is that the districts vary in size substantially. These two factors mean it is very difficult for the square map to approximate real world geography—it is possible to make either the northern half of southern half very accurate but then the other half will be a bit skewed. If you make the northern half accurate the southern half will have a partial gap at the corner of Frederiksberg. If you make the southern half accurate the city centre districts in the northeast end up being misaligned (relative to real world geography). In the end, we chose to make the northern half more accurate because the city centre is located here.

BRØ   Brønshøj-Husum
BIS   Bispebjerg
ØST   Østerbro
VAN   Vanløse
NØR   Nørrebro
IND   Indre By
VAL   Valby
VES   Vesterbro/Kongens Enghave
AMV   Amager Vest
AMØ   Amager Øst

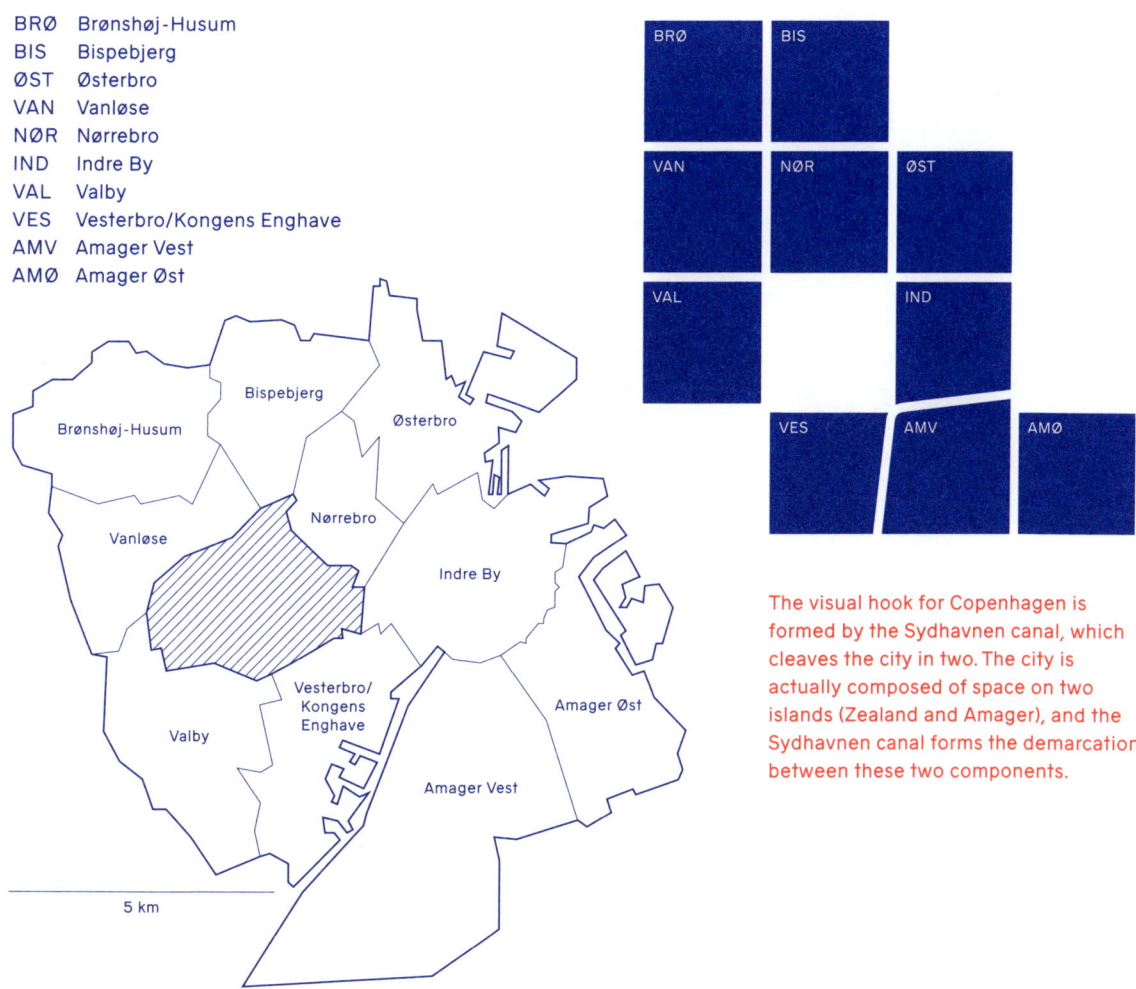

The visual hook for Copenhagen is formed by the Sydhavnen canal, which cleaves the city in two. The city is actually composed of space on two islands (Zealand and Amager), and the Sydhavnen canal forms the demarcation between these two components.

5 km

The area of Copenhagen is of 88 km², with a population of 602,481 people. The strait of Øresund lies to the east. Øresund, strait of water that separates Denmark from Sweden. It passes in between three districts.

35 $88 km^2$

MUMBAI

With its highly irregular shape—an elongated oblong with a wide top and tapering, narrow bottom—Mumbai presents several challenges to forming a sensible squared map. The narrow bottom was actually straightforward to work with, because the districts get smaller as the city outline gets narrower. Additionally, the districts line up neatly.

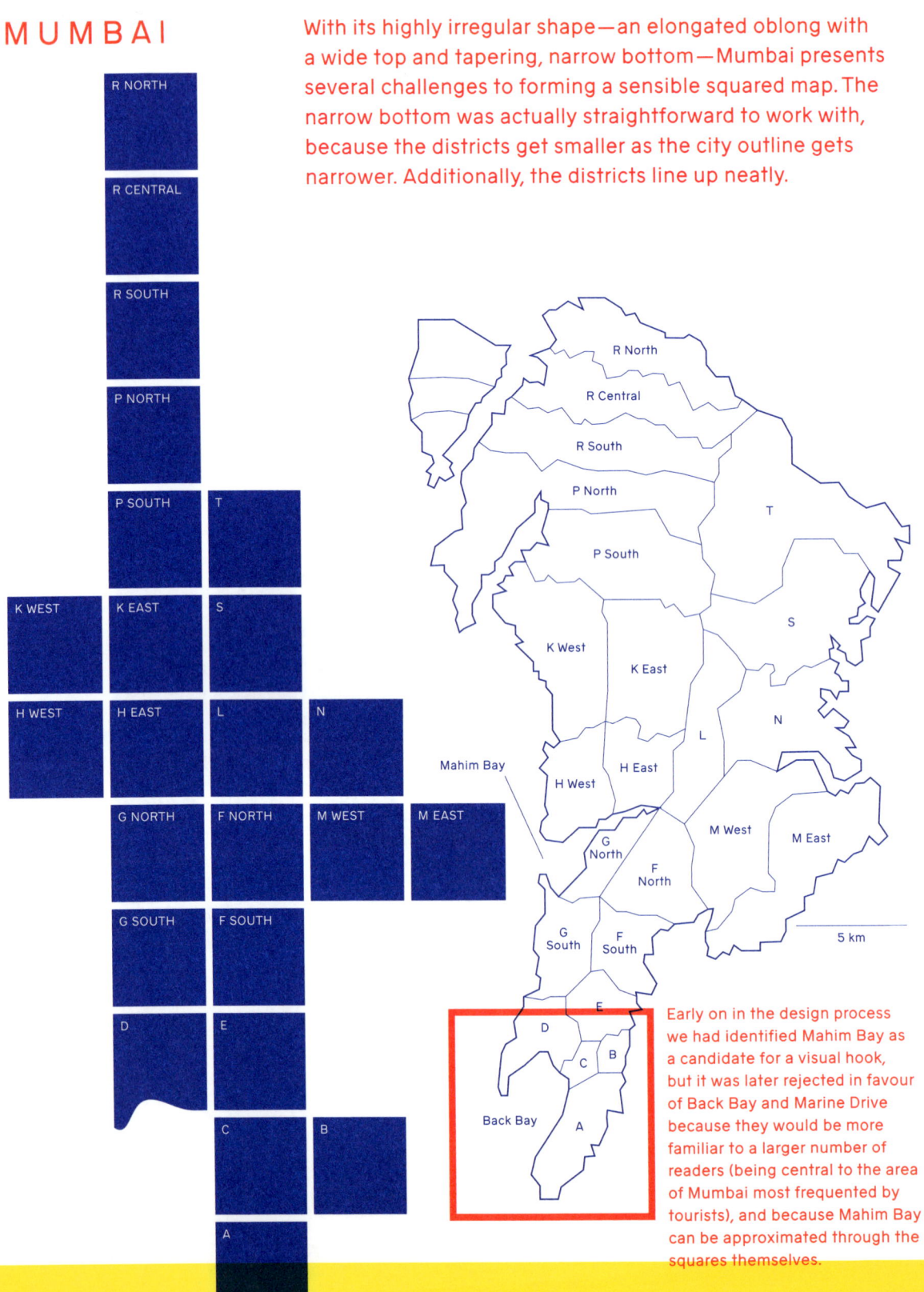

Early on in the design process we had identified Mahim Bay as a candidate for a visual hook, but it was later rejected in favour of Back Bay and Marine Drive because they would be more familiar to a larger number of readers (being central to the area of Mumbai most frequented by tourists), and because Mahim Bay can be approximated through the squares themselves.

The area of Mumbai is of 603 km², with a population of 12,442,373 people. Mumbai is divided into 24 wards and consists of both Mumbai City and the Mumbai Suburban districts.

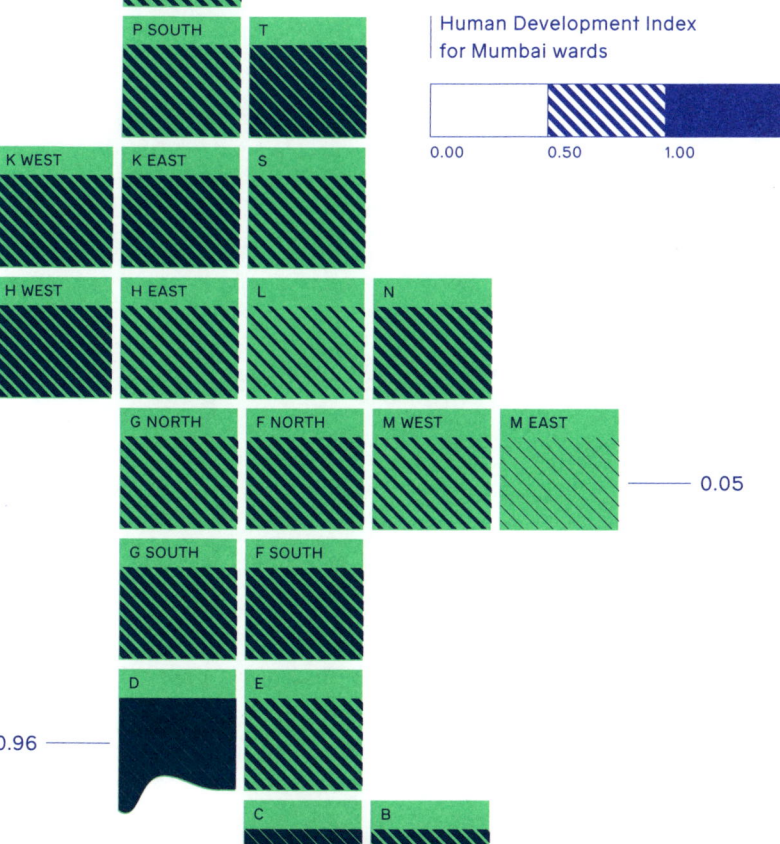

The upper half of the city, by contrast, proved quite challenging. The districts here are very irregular shapes, which are very difficult to arrange in a way that preserves the overall geographic outline of the city. We had to compromise. We stacked the districts vertically, which works well to an extent—but it means that the P and R districts don't connect with T in the way they should. Ultimately, in the square map paradigm, there is simply no way to connect the districts of Mumbai in a way that accurately matches reality.

The Human Development Index is a composite rating that takes into account life expectancy, education, and per capita income. A higher value (closer to 1.00) indicates a better score on the index. Mumbai is an interesting city to examine HDI scores with because it captures almost the entire range of possible scores in a single metropolitan region.

603 km²

37

DÜSSELDORF

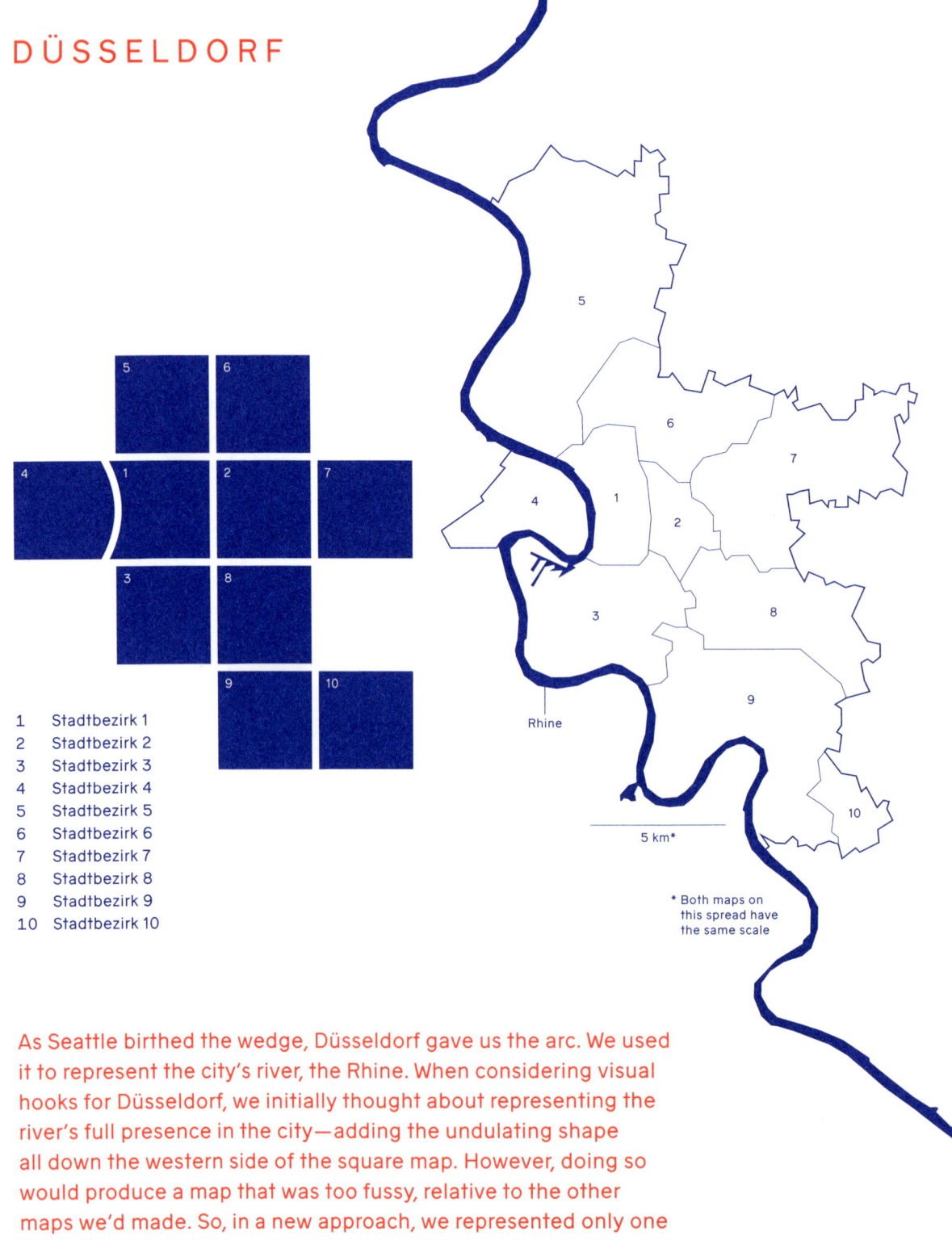

1 Stadtbezirk 1
2 Stadtbezirk 2
3 Stadtbezirk 3
4 Stadtbezirk 4
5 Stadtbezirk 5
6 Stadtbezirk 6
7 Stadtbezirk 7
8 Stadtbezirk 8
9 Stadtbezirk 9
10 Stadtbezirk 10

Rhine

5 km*

* Both maps on this spread have the same scale

As Seattle birthed the wedge, Düsseldorf gave us the arc. We used it to represent the city's river, the Rhine. When considering visual hooks for Düsseldorf, we initially thought about representing the river's full presence in the city—adding the undulating shape all down the western side of the square map. However, doing so would produce a map that was too fussy, relative to the other maps we'd made. So, in a new approach, we represented only one portion of the river, using an arc. This allowed us to create a nice bit of wayfinding that should help a reader spot the centre of the city without distorting the map cells too much.

The area of Düsseldorf is of 217 km², with a population of 613,230 people. The City of Düsseldorf consists of 50 city parts *(Stadtteile),* which are collected into 10 city districts *(Stadtezirke).*

AMSTERDAM

Amsterdam's Noord district is separated from the rest of the city by water, and forms a distinctive protrusion. As such, thoughtful representation of this district provides a strong visual hook for a square map of Amsterdam. The arc represents this district well. Westpoort presented a challenge: it is directly north of Nieuw-West, but its map square can't be placed there without disconnecting Nieuw-West from its eastern neighbour (Zuid). We ruled out placing Westpoort above West because it implied Westpoort and Noord are both North of a dividing river (they are not). Ultimately we chose the geographic distortion as an acceptable trade-off to keep the districts south of the river grouped together as a cluster.

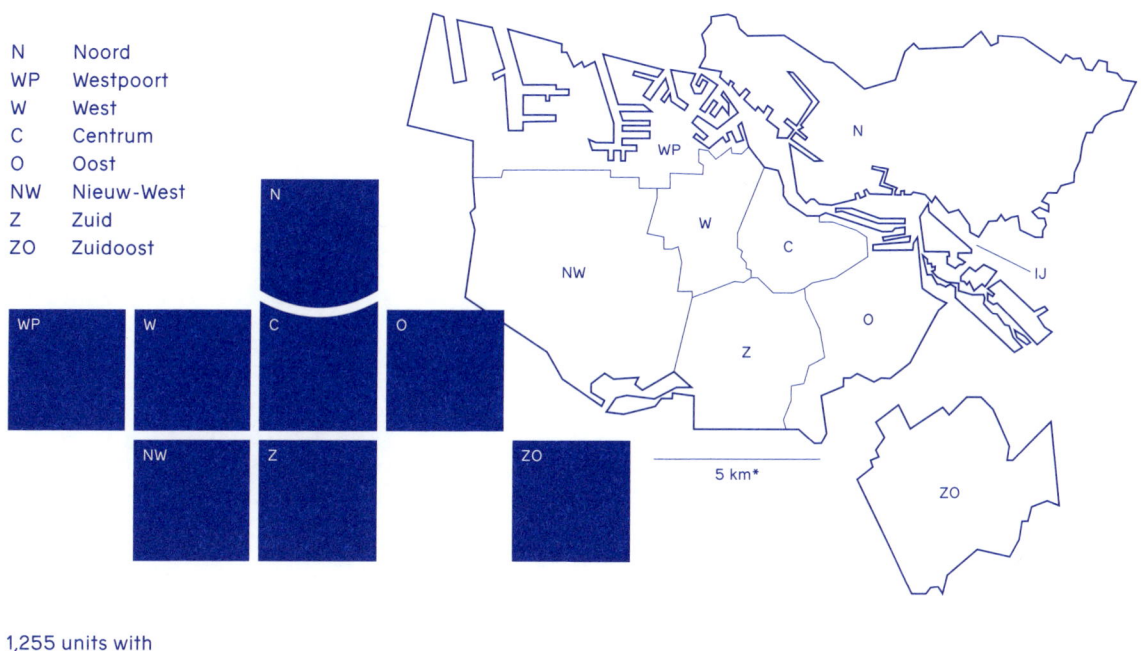

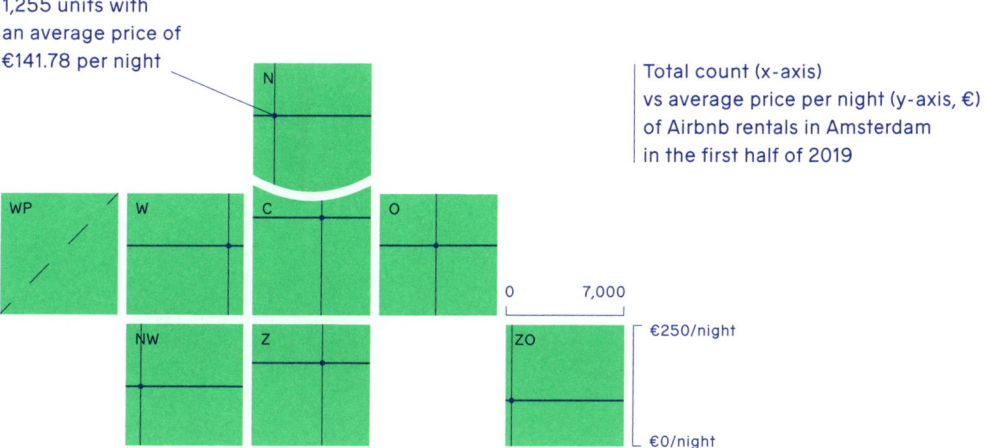

N	Noord
WP	Westpoort
W	West
C	Centrum
O	Oost
NW	Nieuw-West
Z	Zuid
ZO	Zuidoost

5 km*

1,255 units with an average price of €141.78 per night

Total count (x-axis) vs average price per night (y-axis, €) of Airbnb rentals in Amsterdam in the first half of 2019

0   7,000

€250/night

€0/night

The area of Amsterdam is of 219 km², with a population of 851,573 people. The city has eight boroughs, each with a population of around 80,000 to 140,000, which is equivalent to an average-sized municipality in the Netherlands.

219 km²

HELSINKI

Helsinki's district arrangement is distinctive enough that it almost doesn't require any further intervention in the form of a visual hook. The bridge-like arrangement of the square cells creates a very interesting and unique shape. Ultimately, we did add a small wedge shape between the districts of Keskinen-Mellersta and Eteläinen-Södra to approximate the inlet to Töölö Bay that divides the downtown area from the rest of the city.

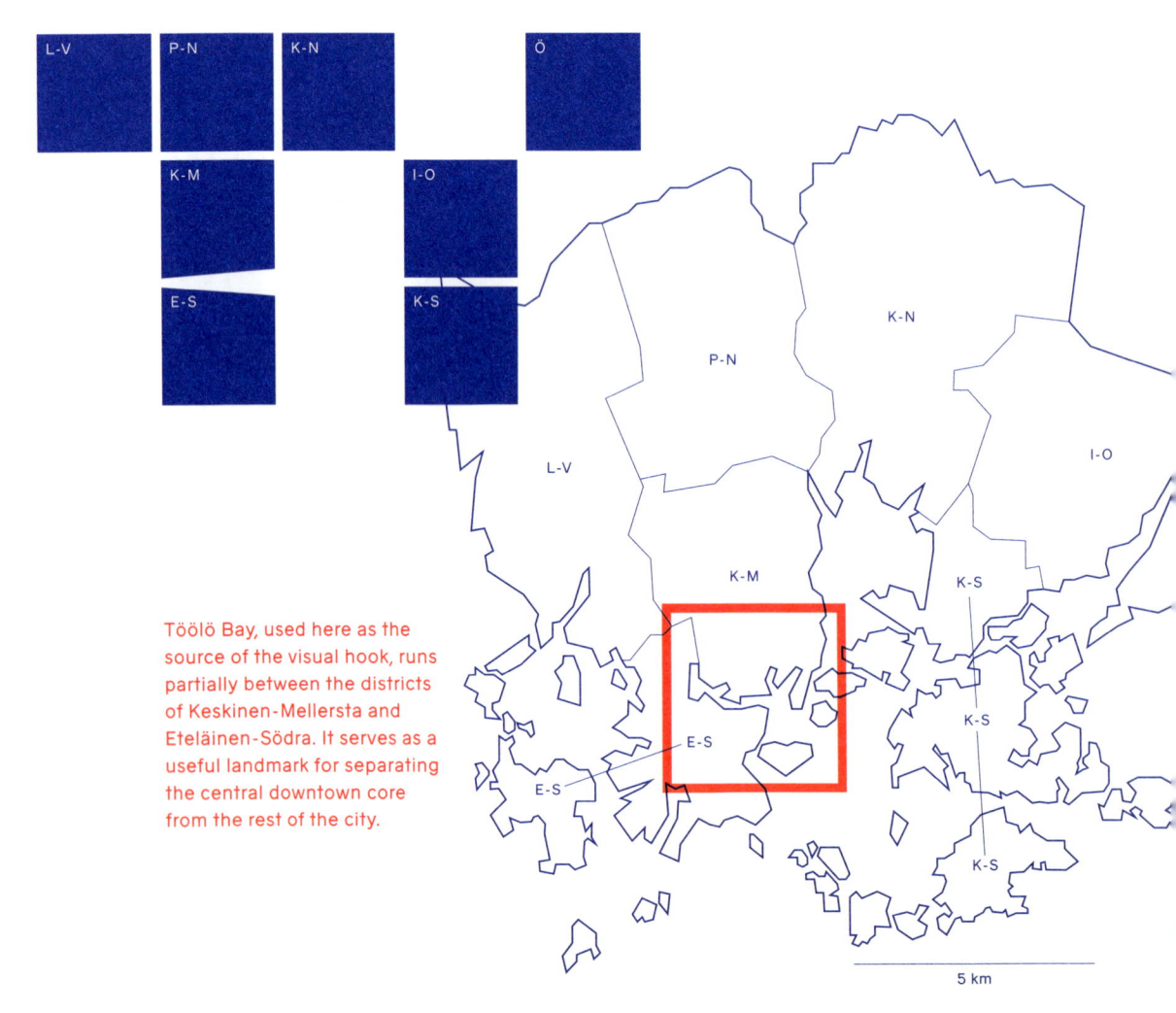

Töölö Bay, used here as the source of the visual hook, runs partially between the districts of Keskinen-Mellersta and Eteläinen-Södra. It serves as a useful landmark for separating the central downtown core from the rest of the city.

The area of Helsinki is of 214 km², with a population of 631,695 people. There are a number of official ways to divide the city, which seems to be a source of some confusion to residents, as different subdivision schemes often share similar or identical names. In addition to the neighbourhood divisions, which exist for city planning purposes, the city is also divided into 34 districts *(Peruspiiri, Distrikt)* to facilitate the coordination of public services. The districts, which may comprise several neighbourhoods, are organised into seven major districts *(Suurpiiri, Stordistrikt)*.

40

L-V   Läntinen-Västra
P-N   Pohjoinen-Norra
K-N   Koillinen-Nordöstra
Ö   Östersundom
K-M   Keskinen-Mellersta
I-O   Itäinen-Östra
E-S   Eteläinen-Södra
K-S   Kaakkoinen-Sydöstra

| Population change, with projections, of Helsinki's major districts

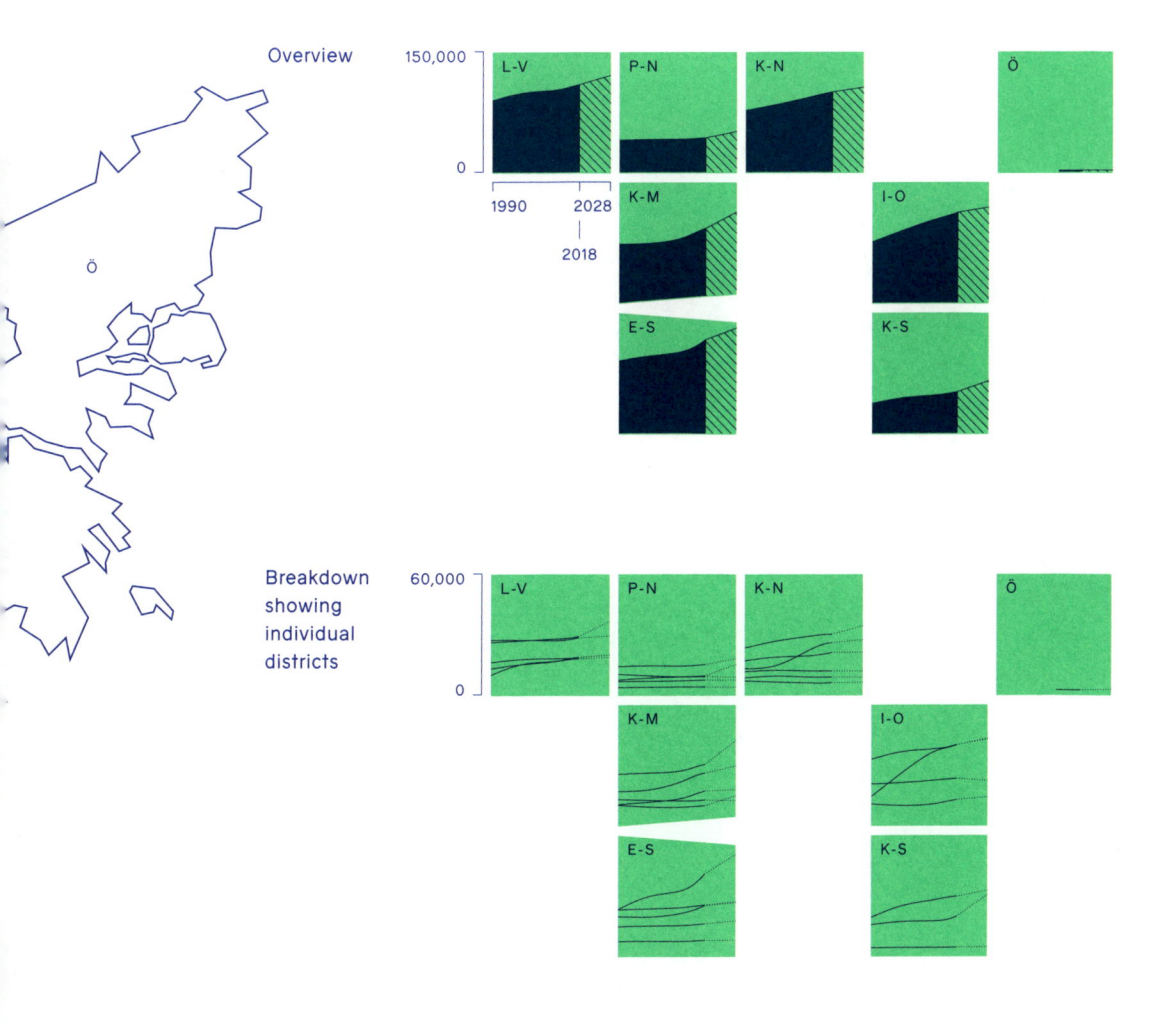

Overview

Breakdown showing individual districts

214 km²

41

AUCKLAND

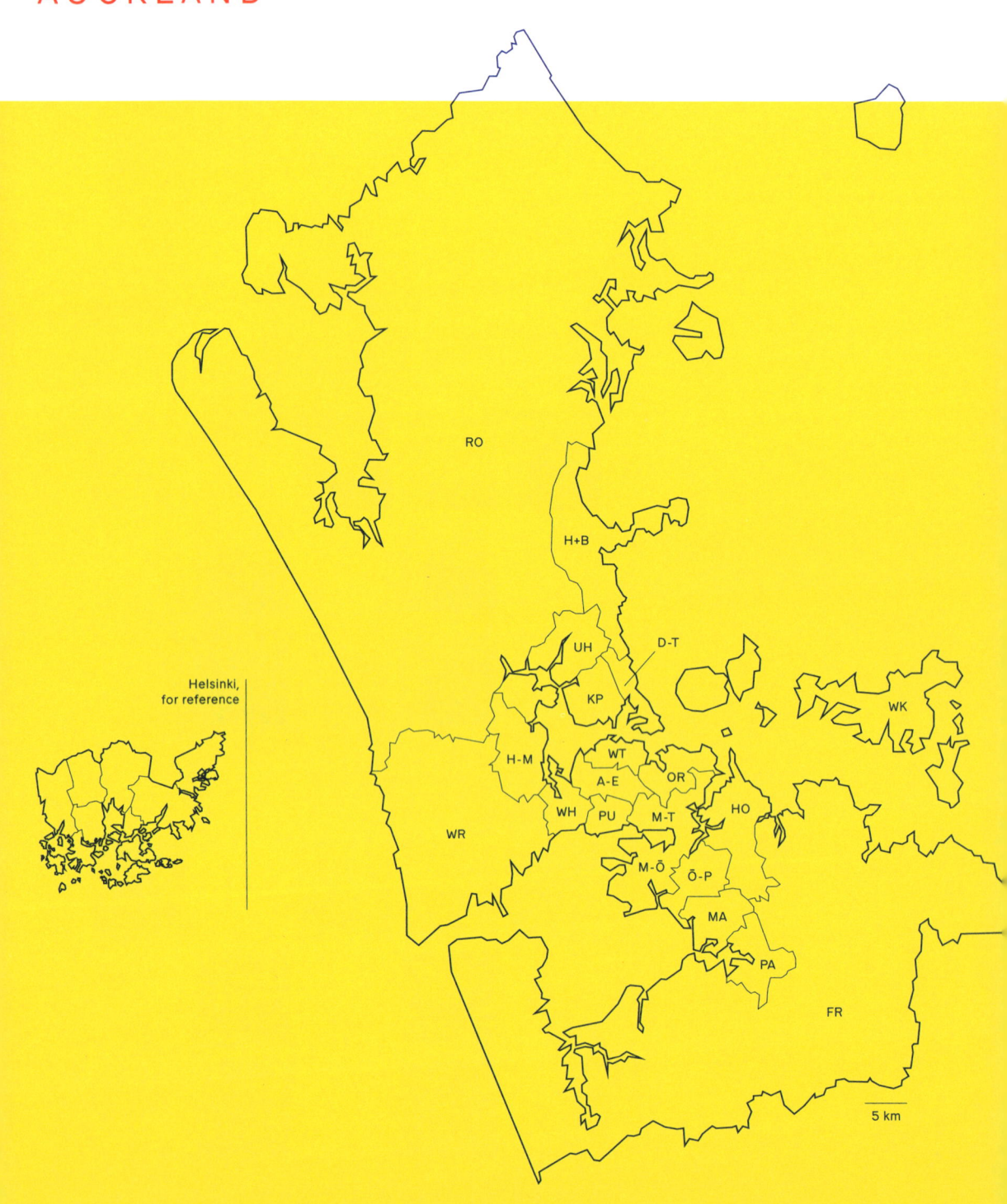

Helsinki, for reference

RO
H+B
UH
D-T
KP
H-M
WT
A-E
OR
WH PU M-T
WR
HO
WK
M-Ō
Ō-P
MA
PA
FR

5 km

The area of Auckland Region is 4,894 km², with a population of 1,695,900 people. The city of Auckland, which is governed by the Auckland Council and part of the greater Auckland Region, is considerably smaller (approximately four times smaller, geographically) and more closely conforms to what a tourist would understand to be the city as such. We've chosen to make a map based on the Auckland Region because, while it is possible to find a map of the metro region, this does not appear to correspond to any standard form of administrative governance and as such would cause Auckland to fall out of step, methodologically, with the other cities in this book.

Auckland's districts provide multiple challenges: they vary greatly in size and shape, they occasionally wrap around one another, and they span several bodies of water. Creating a visual hook was the least of our worries here! More than any other city, this large and complicated city highlighted to key advantage and disadvantage of the square map—the normalisation of cell size makes it easier to compare territories of differing size (here, allowing us to compare small central districts with massive outer districts) but can also distort a sense of real-world geography in extreme ways.

4,894 km²

RO	Rodney
H+B	Hibiscus and Bays
UH	Upper Harbour
KP	Kaipātiki
D-T	Devonport-Takapuna
H-M	Henderson Massey
WA	Waitematā
OR	Orakei
WK	Waiheke
WR	Waitakere Ranges
WH	Whau
A-E	Albert-Eden
M-T	Maungakiekie-Tāmaki
HO	Howick
PU	Puketāpapa
M-Ō	Māngere-Ōtāhuhu
Ō-P	Ōtara-Papatoetoe
MA	Manurewa
PA	Papakura
FR	Franklin

43

HAVANA

As with Cape Town, the best visual hook for a square map of Havana is not immediately obvious when looking at its topographic map. Initially Havana Port seems to be a strong candidate. However, we found we could approximate this shape by arranging the squares in a particular way. So, we made our intervention elsewhere. We used the Malecón—the curving seaside esplanade built on a seawall, whose shape is so distinctive as the visual hook.

PLA	Playa
PR	Plaza de la Revolución
CH	Centro Habana
HV	La Habana Vieja
HdE	La Habana del Este
LIS	La Lisa
MAR	Marianao
CER	Cerro
DdO	Diez de Octubre
REG	Regla
GUA	Guanabacoa
BOY	Boyeros
AN	Arroyo Naranjo
SMP	San Miguel del Padrón
COT	Cotorro

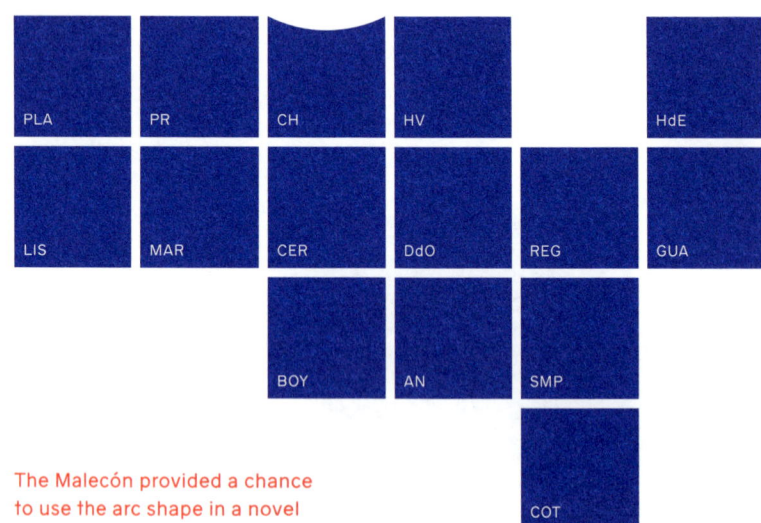

The Malecón provided a chance to use the arc shape in a novel way. In the Havana square map, the arc describes an edge rather than a river curve that divides cells of the map.

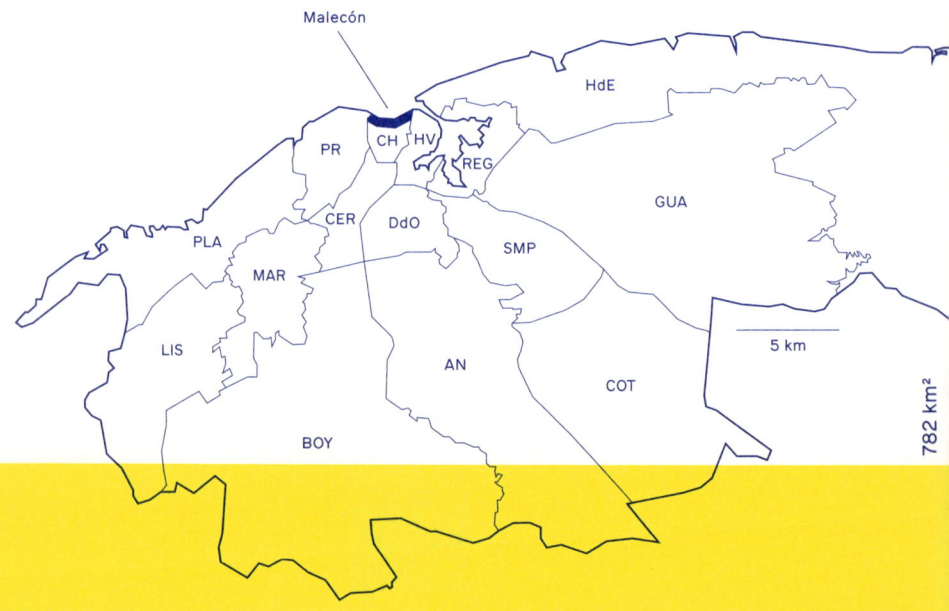

The area of Havana is of 782 km², with a population of 2,106,146 people. The city of Havana is divided into 15 municipalities, or boroughs, which are further subdivided into 105 wards.

CAPE TOWN

When beginning work on a map of Cape Town, the arc of False Bay jumps out as a candidate for a visual hook. However, False Bay this turns out to be a poor wayfinding device, since it is too far from the city centre. Downtown Cape Town, rather than the edge of the city, needed our attention; it is just not recognisable in the square map without intervention in the form of a visual hook. So we created a curve to mimic the western edge of the Western district. This hook it creates a bump-like shape that mirrors the geography of the coast, and highlights the location of the city centre.

Cape Town's square map may contain multiple distortions (in terms of the arrangement of districts), but it has a very strong overall shape. The square map doesn't fully capture the sweeping coastline around False Bay, but it does have a distinct outline with enough of a suggestion of the T-shaped city boundaries.

2,445 km²

WES	Western
NOR	Northern
KLI	Klipfontein
TYG	Tygerberg
SOU	Southern
MIP	Mitchells Plain
KHA	Khayelitsha
EAS	Eastern

The area of The City of Cape Town (Afrikaans: *Stad Kaapstad*; Xhosa: *IsiXeko saseKapa*) is 2,445 km², with a population of 3,740,026 people. The City of Cape Town is divided into eight districts, composed of 115 wards, and is the seat of the Western Cape province.

45

MILAN

Creating a square map of Milan was *very* easy. Its district layout is ideally suited for this kind of project and we didn't need to make any creative interventions.

Daily average download and upload totals (in Megabytes) on public WiFi during the first half of 2019 in Milan districts

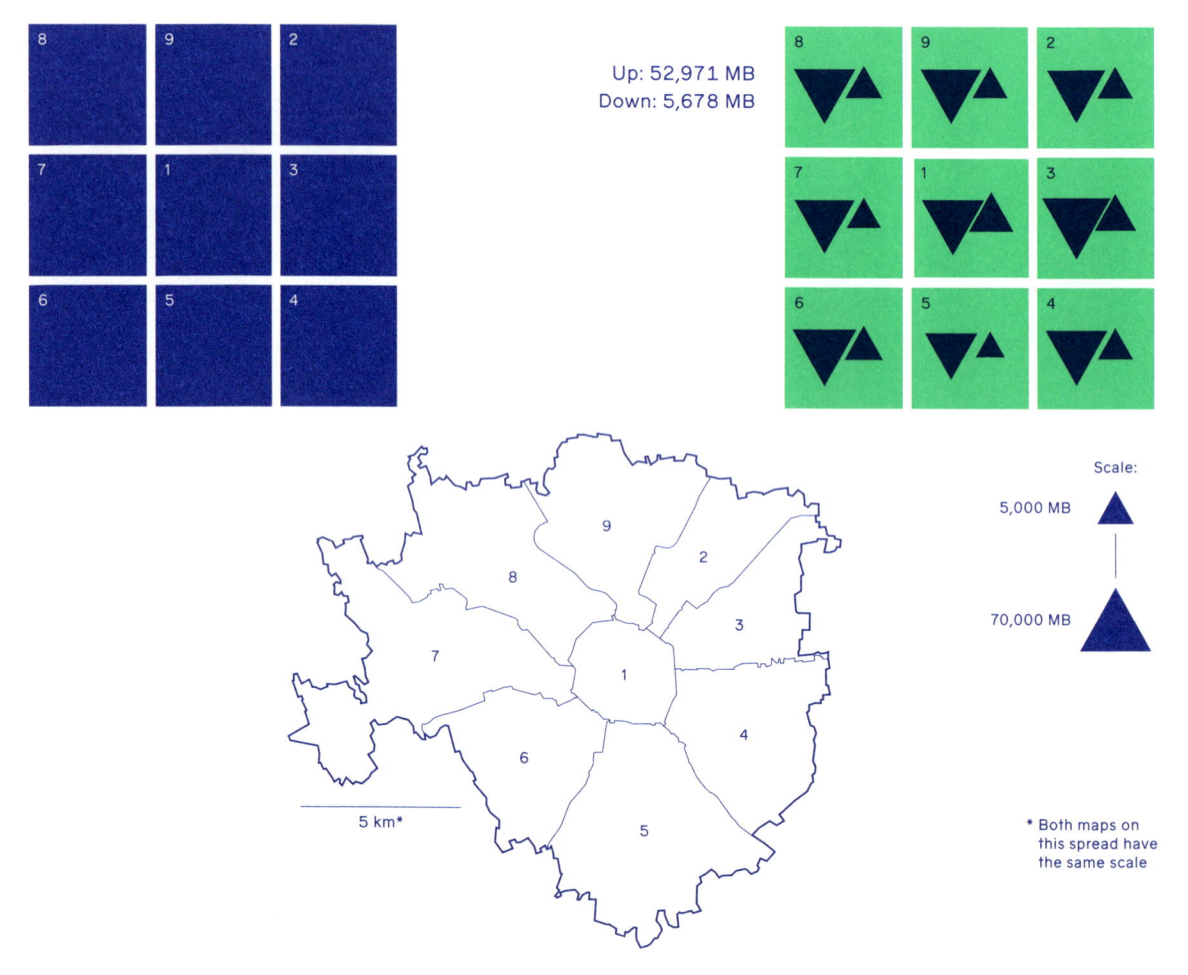

Up: 52,971 MB
Down: 5,678 MB

Scale:

5,000 MB

70,000 MB

5 km*

* Both maps on this spread have the same scale

1	Centro Storico
2	Stazione Centrale, Gorla, Turro, Greco, Crescenzago
3	Città Studi, Lambrate, Porta Venezia
4	Porta Vittoria, Forlanini
5	Vigentino, Chiaravalle, Gratosoglio
6	Barona, Lorenteggio
7	Baggio, De Angeli, San Siro
8	Fiera, Gallaratese, Quarto Oggiaro
9	Porta Garibaldi, Niguarda

The area of Milan is of 182 km², with a population of 1,395,274 people. The city has nine administrative districts, each of which corresponds to a district advisory councils.

46   182 km²

STOCKHOLM

Stockholm is spread across 14 islands in a bay. We used the bay itself—Riddarfjärden—as the visual hook. In order to approximate this bay we used a notch shape. A river shape felt too complicated and wedge shape wasn't expressive enough, and the notch emerged after we tested ways to use two wedges together. The notch is essentially a square rotated 45 degrees. This shape has the advantage of both approximating the shape of Riddarfjärden and clearly marking the city centre. There is some distortion in how the districts are arranged as cells, notably at the extreme north and south. We often observed this kind of distortion at city edges (especially in oblong shaped cities); as the project developed, it became a predictable challenge that we expected to face.

H-V	Hässelby-Vällingby
S-T	Spånga-Tensta
R-K	Rinkeby-Kista
BRO	Bromma
KUN	Kungsholmen
NOR	Norrmalm
ÖST	Östermalm
SKÄ	Skärholmen
H-L	Hägersten-Liljeholmen
SÖD	Södermalm
ÄLV	Älvsjö
EÅV	Enskede-Årsta-Vantör
SPK	Skarpnäck
FAR	Farsta

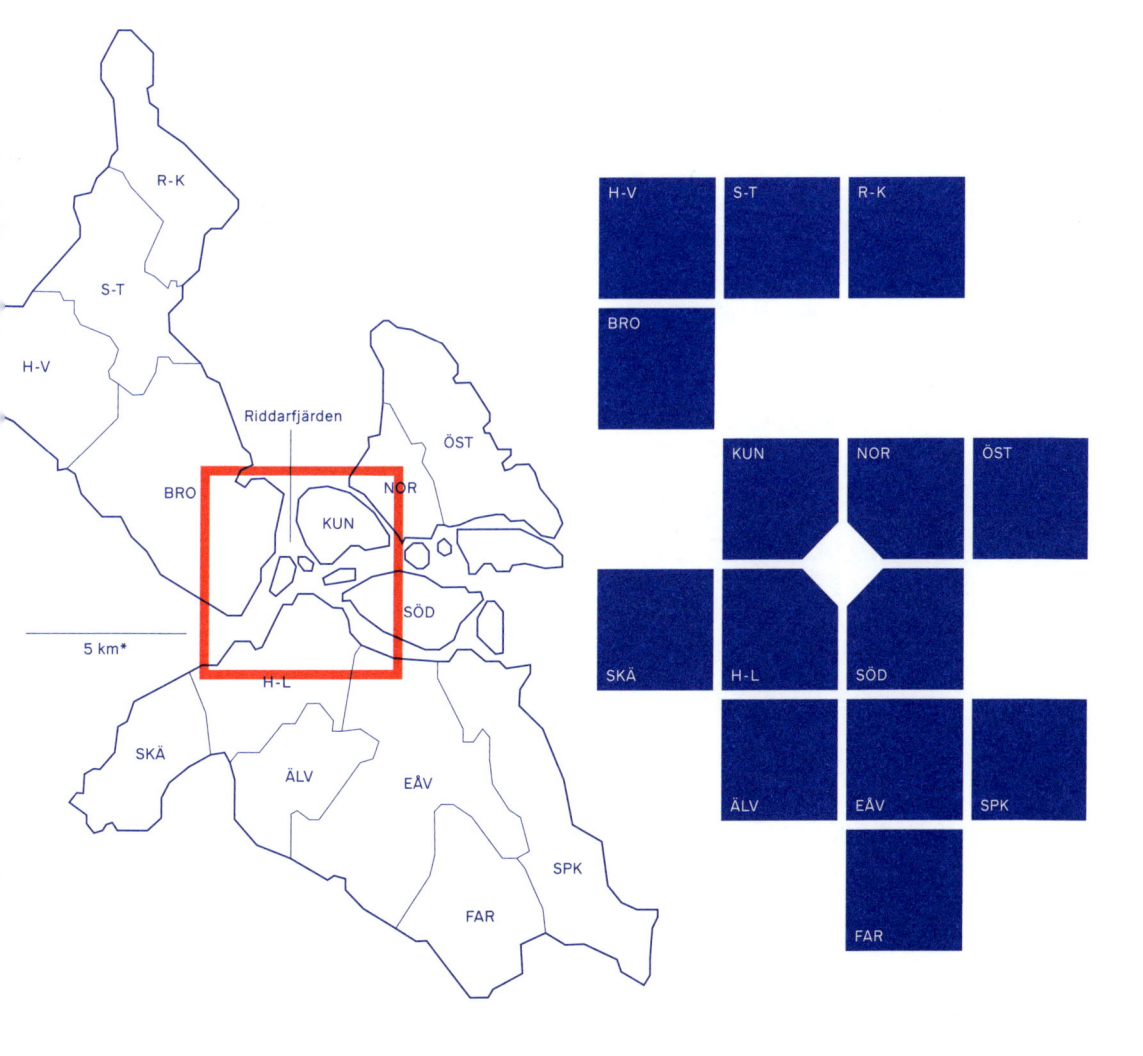

382 km²

The area of Stockholm is of 382 km², with a population of 1,372,565 people. The City of Stockholm is situated on 14 islands amidst a larger archipelago. It has the largest population of the 290 municipalities of the country, but one of the smallest physical areas, making it the most densely populated area of Sweden.

47

KAMPALA

The process of translating the district map of Kampala into a square map was very straightforward, but there were no clear opportunities to add a visual hook in the form of a wedge, arc, etc. Instead the simplicity of the cross formed by the districts becomes its own kind of signature.

Like New York City, Kampala is divided into five districts. Unlike New York City though, Kampala has a relatively small physical footprint and population. This difference is not immediately apparent when looking at the square maps of these cities, side by side. This is because the size of a square map of a city is determined by the number of districts it has—not physical size. While this is useful for looking at data about a single city, it makes side-by-side comparisons between cities more complicated.

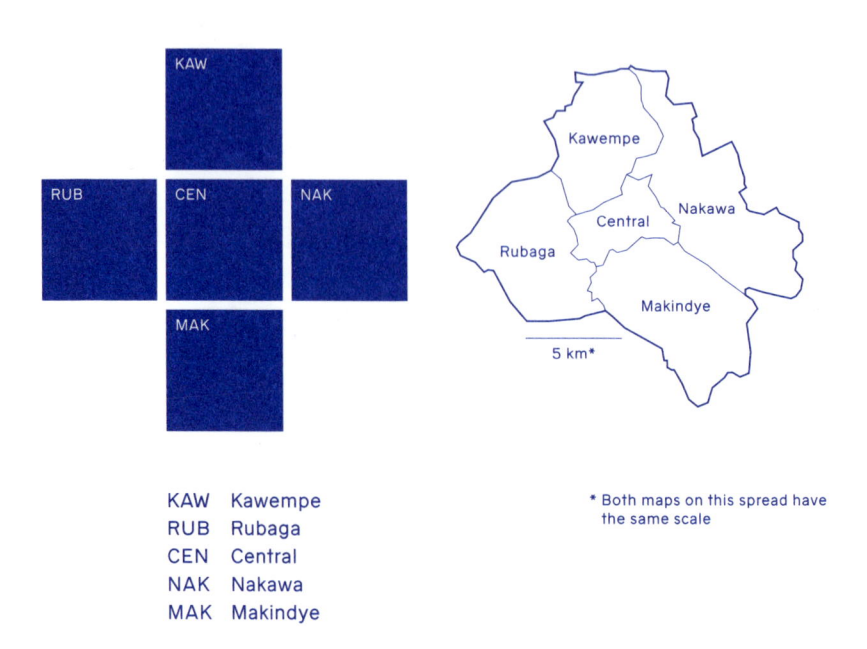

KAW   Kawempe
RUB   Rubaga
CEN   Central
NAK   Nakawa
MAK   Makindye

* Both maps on this spread have the same scale

The area of Kampala is of 189 km², with a population of 1,507,080 people. Kampala is divided into five administrative regions, each headed by a mayor.

48

189 km²

NEW YORK CITY

New York City is a deceptively tricky city to translate into a square map—whilst it is arranged in a tidy grid, some of the boroughs are very small, and the whole grid is actually titled 30° on its north/south axis. We tried arranging the boroughs in a geographically accurate manner (with the Bronx placed North of Queens, Manhattan to the West of Queens, and Brooklyn placed South of Manhattan). However, this arrangement didn't really look or feel like the city. This is mainly because most residents don't recognise that the city grid is tilted, and think of the city in terms of north and south. Re-orienting our map to residents' vertical imagination still didn't completely work though, for various reasons. After a while we settled on a small cheat, of sorts—shifting Manhattan and the Bronx up half a unit so that the left side of the map would be offset against the right. It is a simple move, but feels both accurate and oddly satisfying.

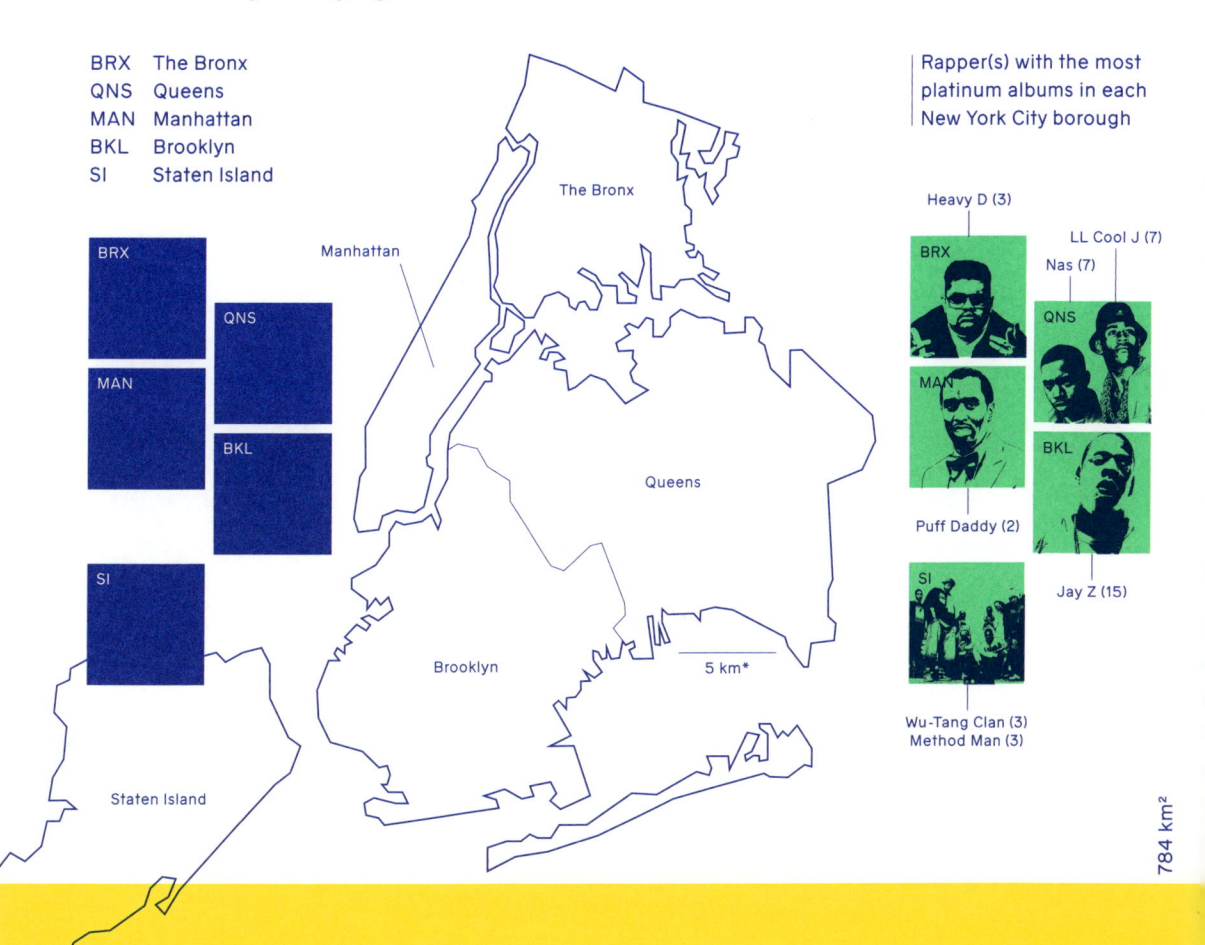

BRX The Bronx
QNS Queens
MAN Manhattan
BKL Brooklyn
SI   Staten Island

Rapper(s) with the most platinum albums in each New York City borough

BRX — Heavy D (3)
QNS — LL Cool J (7), Nas (7)
MAN — Puff Daddy (2)
BKL — Jay Z (15)
SI — Wu-Tang Clan (3), Method Man (3)

5 km*

784 km²

The area of NYC is of 784 km², with a population of 8,398,748 people. New York City encompasses five county-level administrative divisions called boroughs: The Bronx, Brooklyn, Manhattan, Queens, and Staten Island.

TOKYO

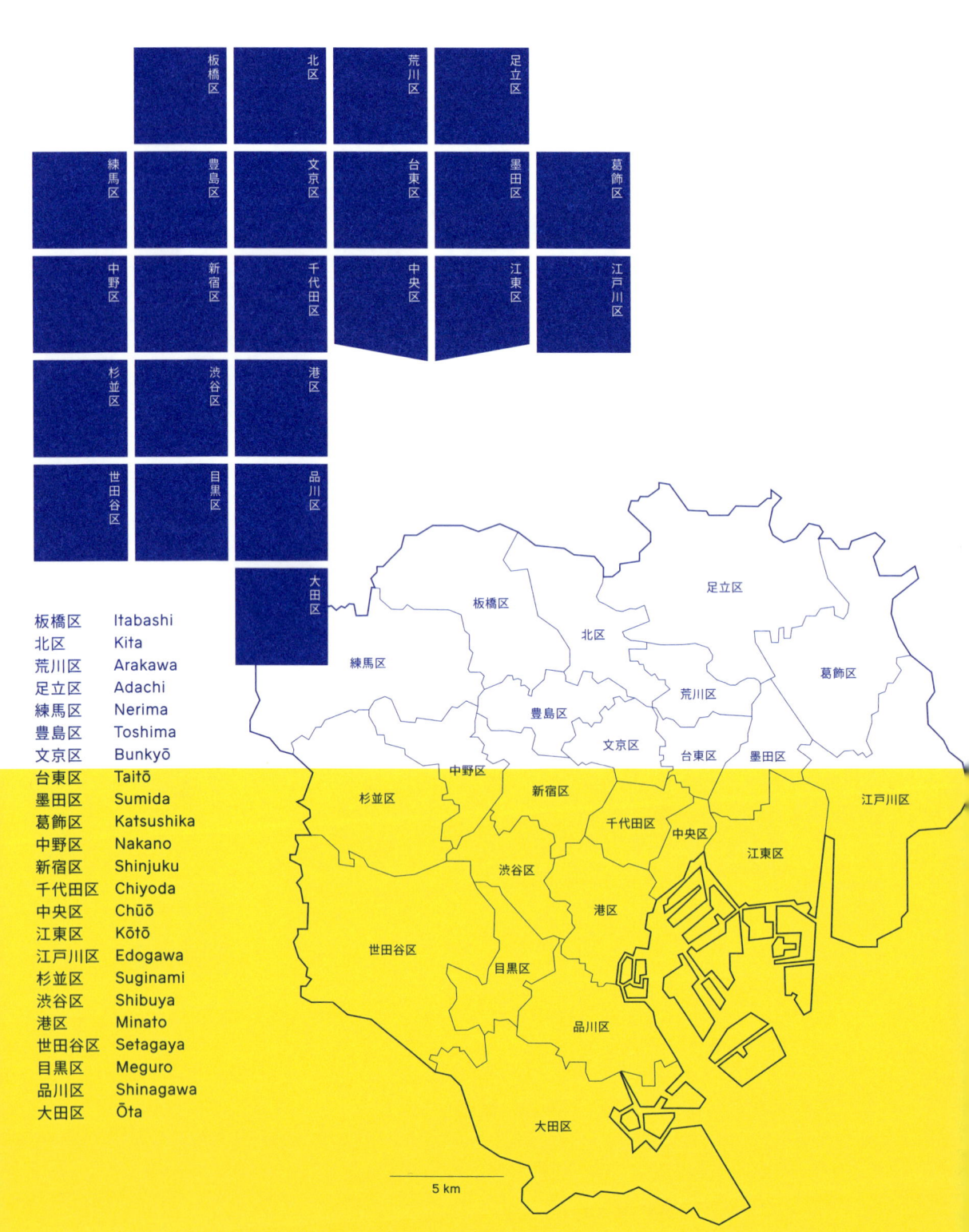

板橋区	Itabashi
北区	Kita
荒川区	Arakawa
足立区	Adachi
練馬区	Nerima
豊島区	Toshima
文京区	Bunkyō
台東区	Taitō
墨田区	Sumida
葛飾区	Katsushika
中野区	Nakano
新宿区	Shinjuku
千代田区	Chiyoda
中央区	Chūō
江東区	Kōtō
江戸川区	Edogawa
杉並区	Suginami
渋谷区	Shibuya
港区	Minato
世田谷区	Setagaya
目黒区	Meguro
品川区	Shinagawa
大田区	Ōta

The area of Tokyo is of 2,188 km², with a population of 38,140,000 people. The core and most populous part of Tokyo is comprised of 23 districts called "Special Wards". These wards comprise the original Tokyo City before it was subsumed in 1943 to become part of the newly created Tokyo Metropolis.

50

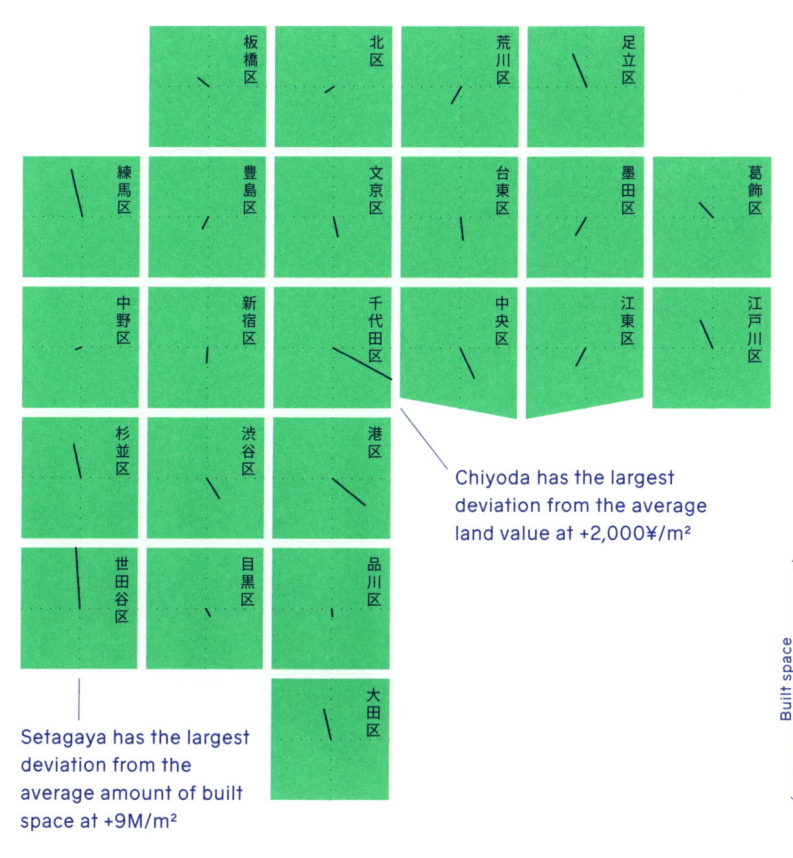

Deviation from the mean for residential land value (x-axis) vs total built space (y-axis) in Tokyo Special Wards

板橋区	北区	荒川区	足立区
練馬区	豊島区	文京区	台東区	墨田区	葛飾区
中野区	新宿区	千代田区	中央区	江東区	江戸川区
杉並区	渋谷区	港区
世田谷区	目黒区	品川区
		大田区

Chiyoda has the largest deviation from the average land value at +2,000¥/m²

Setagaya has the largest deviation from the average amount of built space at +9M/m²

The x-axis shows how far a ward varies from the average residential land value. The longest line along this axis represents a deviation of plus or minus 2,000¥ per m².

The y-axis shows how far a ward varies from the average total built space. The longest line along this axis represents a deviation of plus or minus 9,000,000 m².

	Value	
	−2k ¥/m²　mean　+2k ¥/m²	
+9M/m²	low value high space	high value high space
mean		
−9M/m²	low value low space	high value low space

Built space

2,188 km²

Creating a visual hook for a square map of Tokyo was challenging. We considered using the large Arakawa River, but it passes through wards, making this option unviable. We also tried to use the smaller Sumida River, but it is difficult to render the river without including too many curves, or to keep it one line without breaks (as the river must jump around squares). In the end, instead of using the rivers as hooks, we chose to use the city's port as a visual hook. We represent it with a sort of beak form, which evokes the shape of the docklands where the districts of Chūō and Kōtō meet.

51

TAIPEI

Tapei's curving Tamsui River is the obvious candidate for a visual hook. Figuring out how to implement the shape, and how to arrange the city's districts, was less obvious. The districts of Neihu and Shilin are the troublemakers. Both connect with multiple districts on one of their edges, and so the shape of the square map diverges considerably from the geographic map. Many compromises are required to make a square map of Taipei. But, in the end, these compromises don't undermine the success of the map. Taipei showed us how far we could push the concept, in a city where the geography resists our methods.

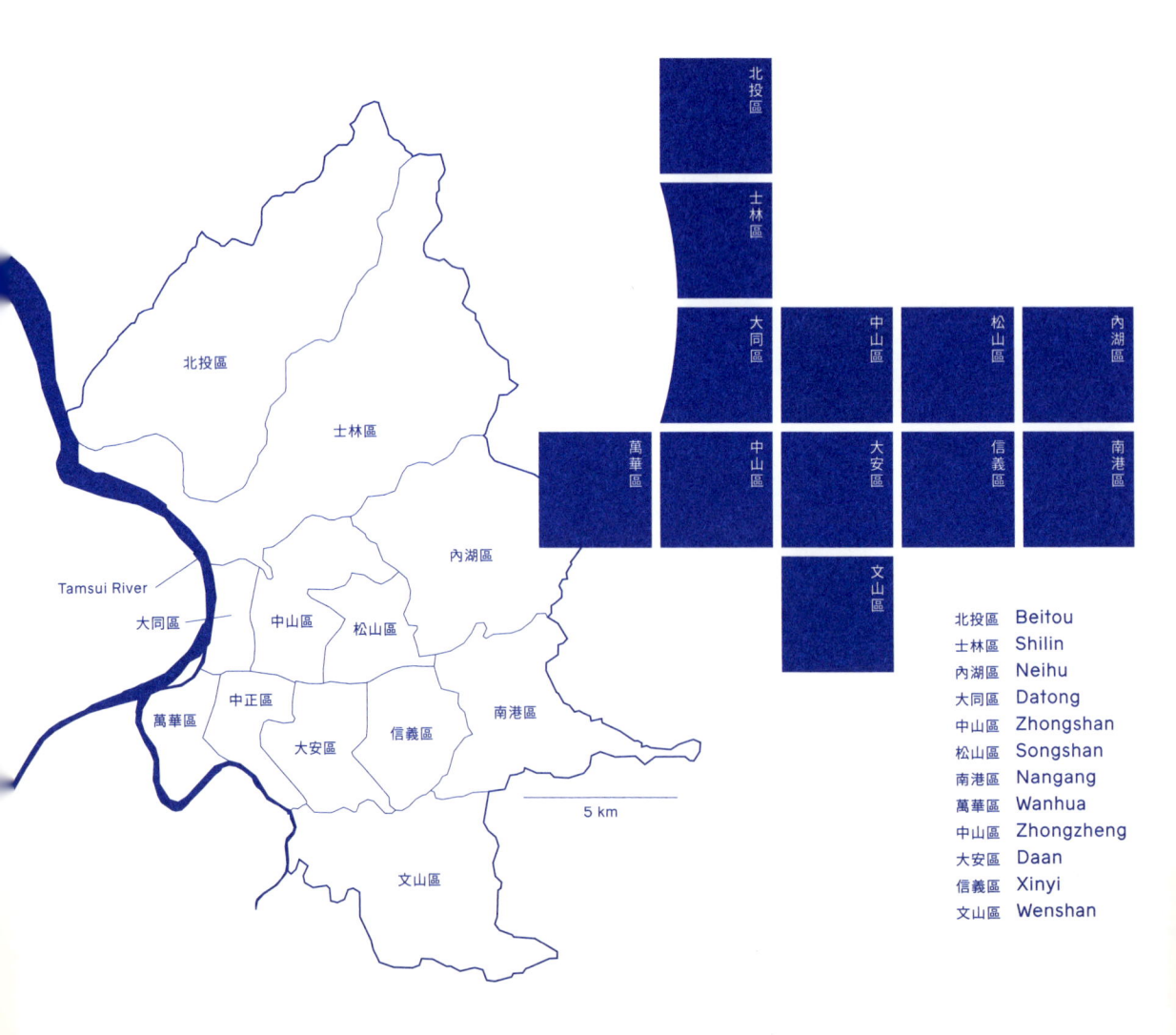

北投區	Beitou
士林區	Shilin
內湖區	Neihu
大同區	Datong
中山區	Zhongshan
松山區	Songshan
南港區	Nangang
萬華區	Wanhua
中山區	Zhongzheng
大安區	Daan
信義區	Xinyi
文山區	Wenshan

272 km²

The area of Taipei is 272 km², with a population of 2,674,063. The city is divided into 12 administrative districts. While it is easily the most dense city in Taiwan it is only the fourth most populated.

LAGOS

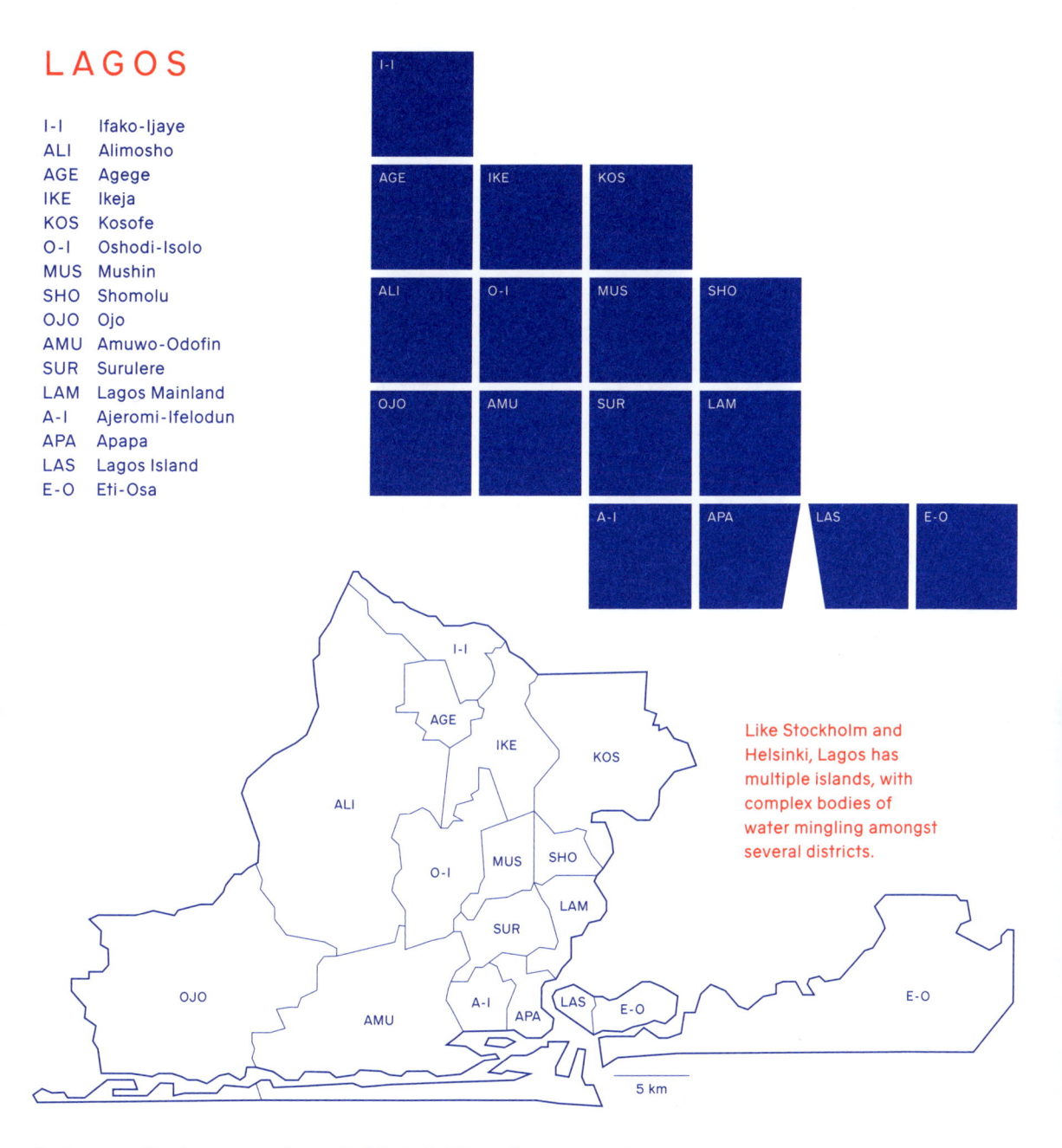

I-I	Ifako-Ijaye
ALI	Alimosho
AGE	Agege
IKE	Ikeja
KOS	Kosofe
O-I	Oshodi-Isolo
MUS	Mushin
SHO	Shomolu
OJO	Ojo
AMU	Amuwo-Odofin
SUR	Surulere
LAM	Lagos Mainland
A-I	Ajeromi-Ifelodun
APA	Apapa
LAS	Lagos Island
E-O	Eti-Osa

Like Stockholm and Helsinki, Lagos has multiple islands, with complex bodies of water mingling amongst several districts.

1,171 km²

In Lagos, the large western district of Alimosho presents an immediate challenge when making a square map, since it borders four other districts on its eastern edge alone (see the one-to-several problem, page 23). The overall shape works, but it does tilt in the wrong direction when compared with the geographic map. We used Lagos Lagoon as a visual hook. This waterway separates the districts of Lagos Island and Eti-Osa from the mainland. We introduced a wedge at the bottom of the square map to indicate the entry point of the Lagos Lagoon.

The area of Lagos is of 1,171 km², with a population of 6,048,430 people. Lagos State is divided into five administrative divisions, which are further divided into 20 local government areas, or LGAs.

GLASGOW

MAR   Maryhill
CAN   Canal
S/R   Springburn/Robroyston
NE   North East
D/A   Drumchapel/Anniesland
P/K   Partick East/Kelvindale
HIL   Hillhead
DEN   Dennistoun
EC   East Centre
BAI   Baillieston
G/S   Garscadden/Scotstounhill
VP   Victoria Park
A/C   Anderston/City/Yorkhill
CAL   Calton
SHE   Shettleston
GOV   Govan
SC   Southside Central
CAR   Cardonald
POL   Pollokshields
LAN   Langside
LIN   Linn
GP   Greater Pollok
N/A   Newlands/Auldburn

5 km

The area of Glasgow is 175 km², with a population of 621,020. The city is divided into 23 wards. The current district layout dates to 2017.

54

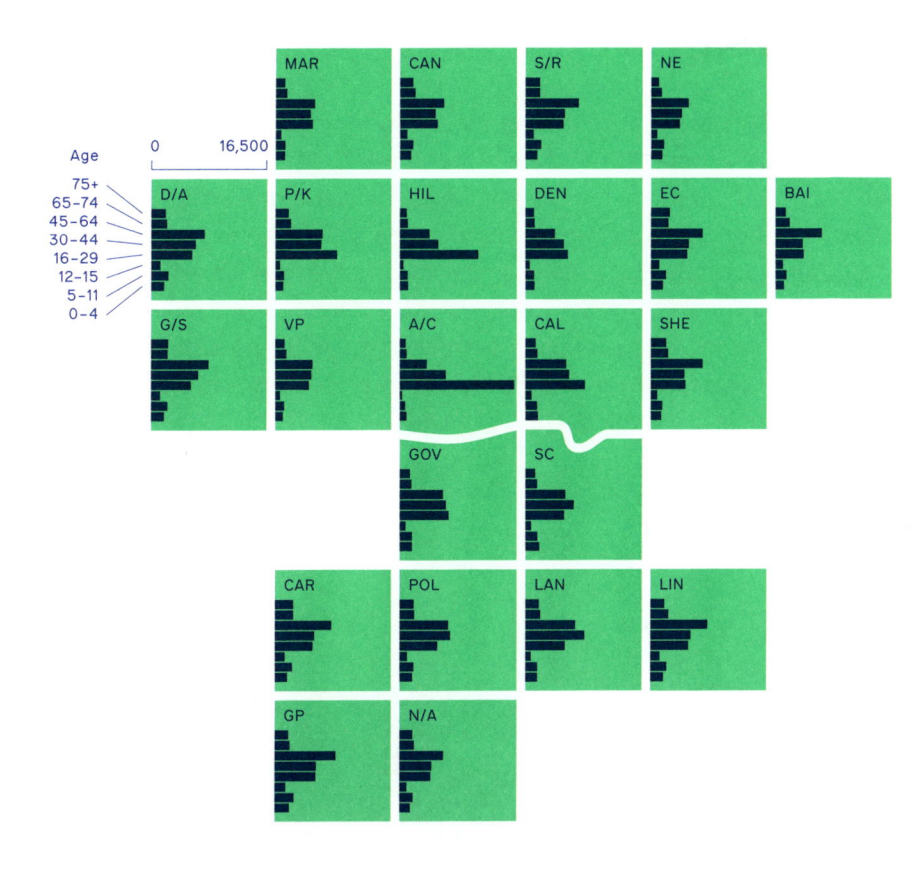

Population (x-axis) by age groups (y-axis) in Glasgow wards

Age: 75+, 65–74, 45–64, 30–44, 16–29, 12–15, 5–11, 0–4

0 — 16,500

Wards: MAR, CAN, S/R, NE, D/A, P/K, HIL, DEN, EC, BAI, G/S, VP, A/C, CAL, SHE, GOV, SC, CAR, POL, LAN, LIN, GP, N/A

Glasgow is notable for the large number of districts it has for its relatively small geographic footprint. Its 23 divisions far outnumber cities of a similar size like Kampala and Milan, which have 5 and 9 divisions, respectively. Like many cities situated on a river, Glasgow is divided in half. The River Clyde creates a simple division between the north and south, with the usefully distinct wiggle between Calton and Southside Central providing the visual hook. Perhaps more interesting than the visual hook for this map is how the translation from geography to map squares exaggerates the shape of the city—the pinched waist below the river is given a greater emphasis in the square map because the ward of Govan needs to be condensed horizontally.

55   175 km²

MOSCOW

The River Moskva snakes through Moscow in a manner that is difficult to manage. We used far more of the river's curves here than we did in the London's map, because a minimal representation does not adequately evoke this serpentine waterway. Moscow's peripheral districts also posed challenges. Whilst the district of Zelenogradsky was relatively easy to place, Novomoskovsky, and Troitsky were more challenging as they sit on an angle. Ultimately we did compromise by placing these squares on a 45° angle to better mimic the overall shape of the city.

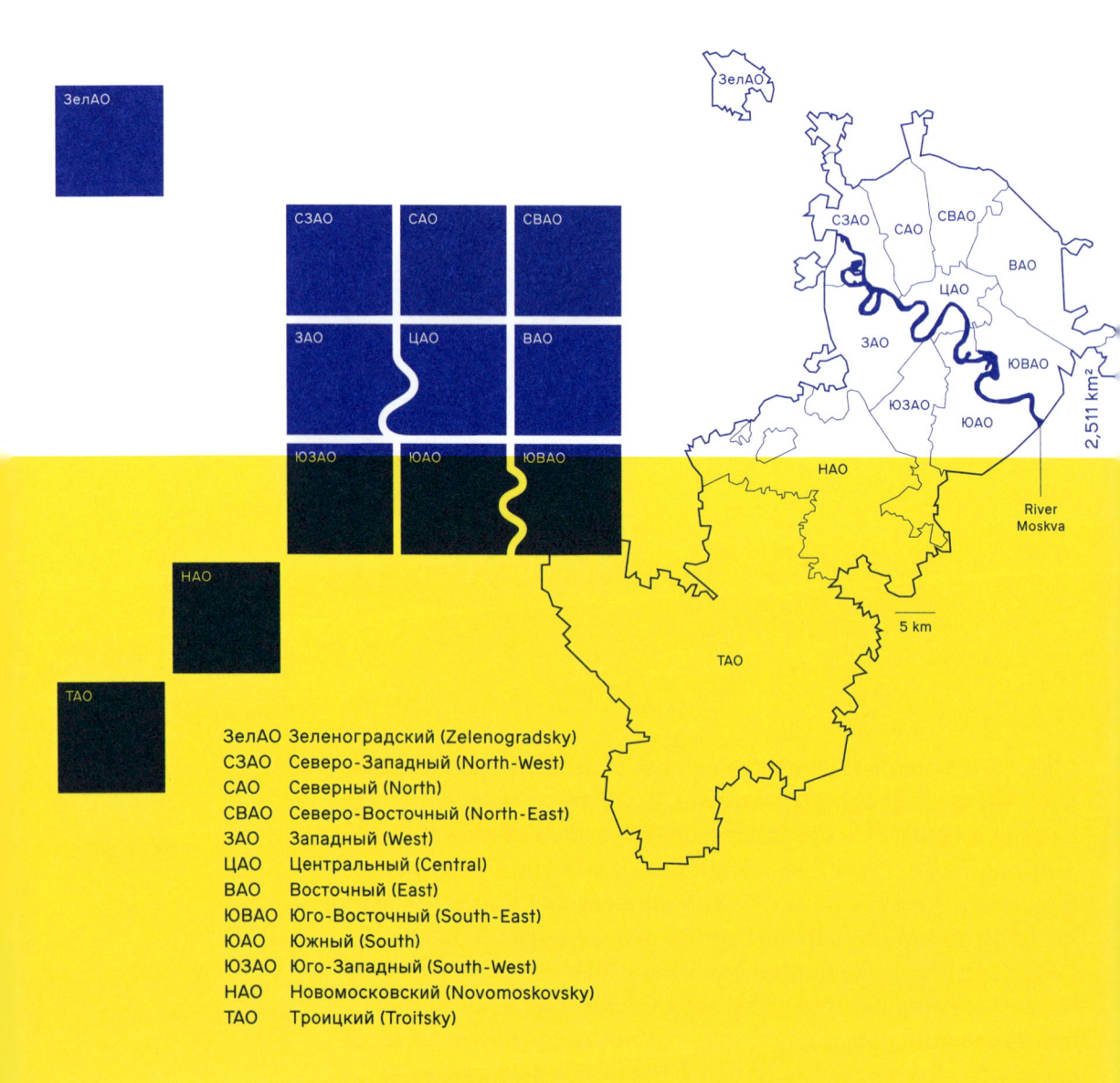

ЗелАО Зеленоградский (Zelenogradsky)
СЗАО Северо-Западный (North-West)
САО Северный (North)
СВАО Северо-Восточный (North-East)
ЗАО Западный (West)
ЦАО Центральный (Central)
ВАО Восточный (East)
ЮВАО Юго-Восточный (South-East)
ЮАО Южный (South)
ЮЗАО Юго-Западный (South-West)
НАО Новомосковский (Novomoskovsky)
ТАО Троицкий (Troitsky)

The area of Moscow is 2,511 km², with a population of 12,506,468 people. Moscow is divided into 12 divisions also known as *okrugs*, which are in turn divided into districts known as *raions*. The current district arrangement dates to 2012, when Novomoskovsky and Troitsky were incorporated into the city of Moscow.

56

REYKJAVÍK

Reykjavík's square map was both very simple and very complicated to devise. The core districts (1–8 and 10) can be laid out with very little trouble, and the inlet that leads to the Elliðaá River is an obvious candidate for a visual hook. However, Kjalarnes complicates matters considerably. Kjalarnes is cut off from the rest of the city by the bay. Whilst we considered detaching the cell from the others, it made it feel like an island, which it isn't. In the end we used the notch shape to show the bay without fully detaching the Kjalarnes map square.

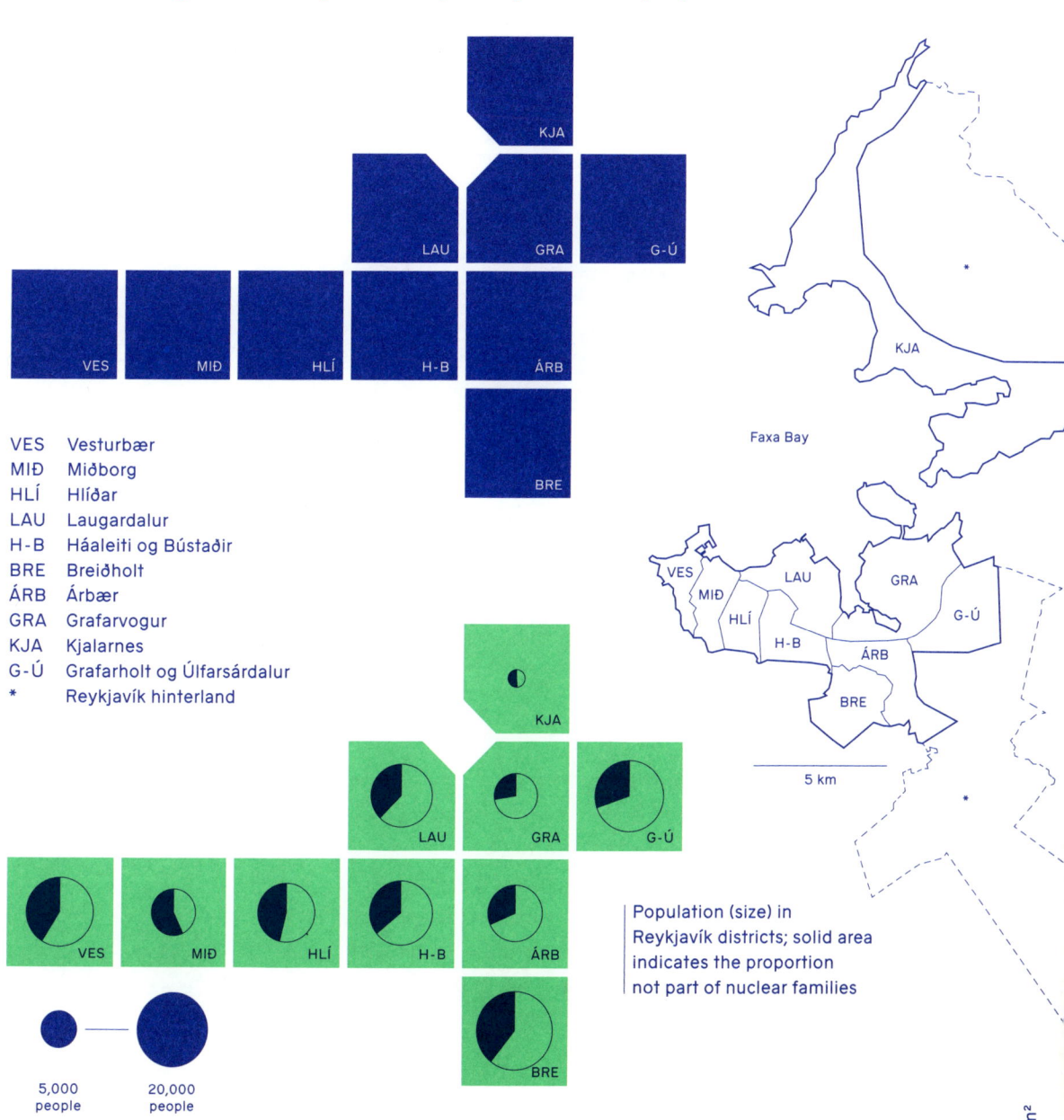

VES — Vesturbær
MIÐ — Miðborg
HLÍ — Hlíðar
LAU — Laugardalur
H-B — Háaleiti og Bústaðir
BRE — Breiðholt
ÁRB — Árbær
GRA — Grafarvogur
KJA — Kjalarnes
G-Ú — Grafarholt og Úlfarsárdalur
* — Reykjavík hinterland

Faxa Bay

5 km

5,000 people
20,000 people

Population (size) in Reykjavík districts; solid area indicates the proportion not part of nuclear families

273 km²

The area of Reykjavík is of 273 km², with a population of 128,793. There are ten districts in Reykjavík, one of which is not contiguous with the other nine. In addition to the ten city districts of Reykjavík there are two hinterland regions that belong to the city but are not assigned to any district.

57

MONTREAL

1	Ahuntsic-Cartierville
2	Anjou
3	Côte-des-Neiges—Notre-Dame-de-Grâce
4	Lachine
5	LaSalle
6	Le Plateau-Mont-Royal
7	Le Sud-Ouest
8	L'Île-Bizard–Sainte-Geneviève
9	Mercier—Hochelaga-Maisonneuve
10	Montréal-Nord
11	Outremont
12	Pierrefonds-Roxboro
13	Rivière-des-Prairies—Pointe-aux-Trembles
14	Rosemont–La Petite-Patrie
15	Saint-Laurent
16	Saint-Léonard
17	Verdun
18	Ville-Marie
19	Villeray–Saint-Michel—Parc-Extension

*Both maps on this spread have the same scale

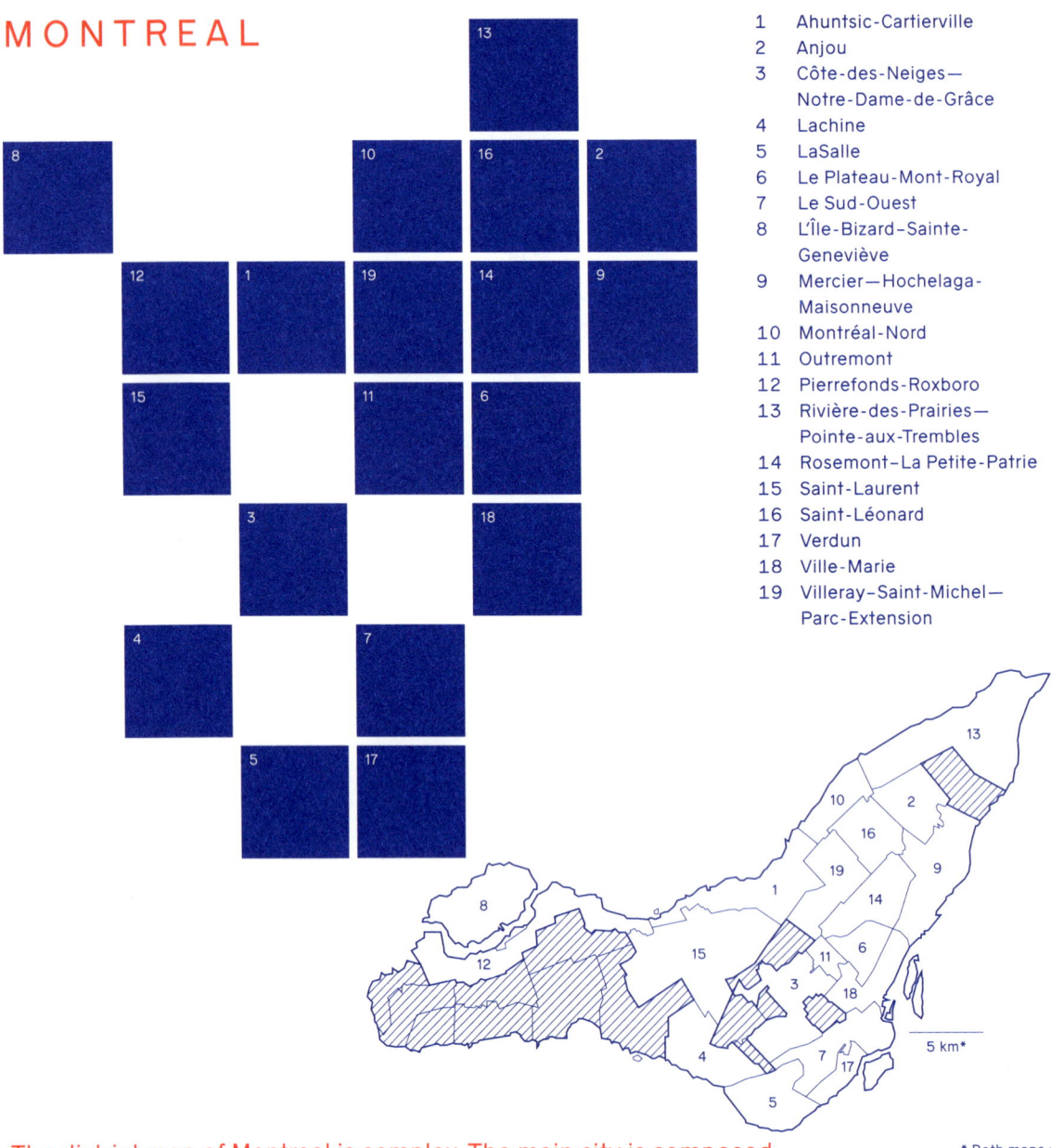

The district map of Montreal is complex. The main city is composed of 19 districts. These are situated on the Island of Montreal—but the island is not synonymous with the city of Montreal. The Island also includes a further 14 unincorporated districts. To complicate matters further, some of these unincorporated districts are completely surrounded by other city districts—so, as with Frederiksberg in Copenhagen, they create gaping holes in our square map. However, all of this complexity creates distinction. The complex shape of Montreal became its own visual hook.

The area of Montreal is of 365 km², with a population of 1,704,694 people. The city has 19 boroughs (*arrondissements*), each with a mayor and council. In 2002 all of the municipalities of Montreal island were merged to create a city of 27 boroughs, however there was a demerger in 2006 resulting in the current 19 borough composition.

365 km²

58

BAGHDAD

The River Tigris splits the city of Baghdad in half. The Tigris' extreme curves create some very irregularly shaped districts, not least Karrada, where the river creates a deep peninsula, with the University of Baghdad on its tip. Making the river recognisable without representing every curve was a challenge. We also faced issues with scale. The eastern side of Baghdad is smaller than the western side, but in the square map this is reversed. This is because the western districts are larger than the eastern districts.

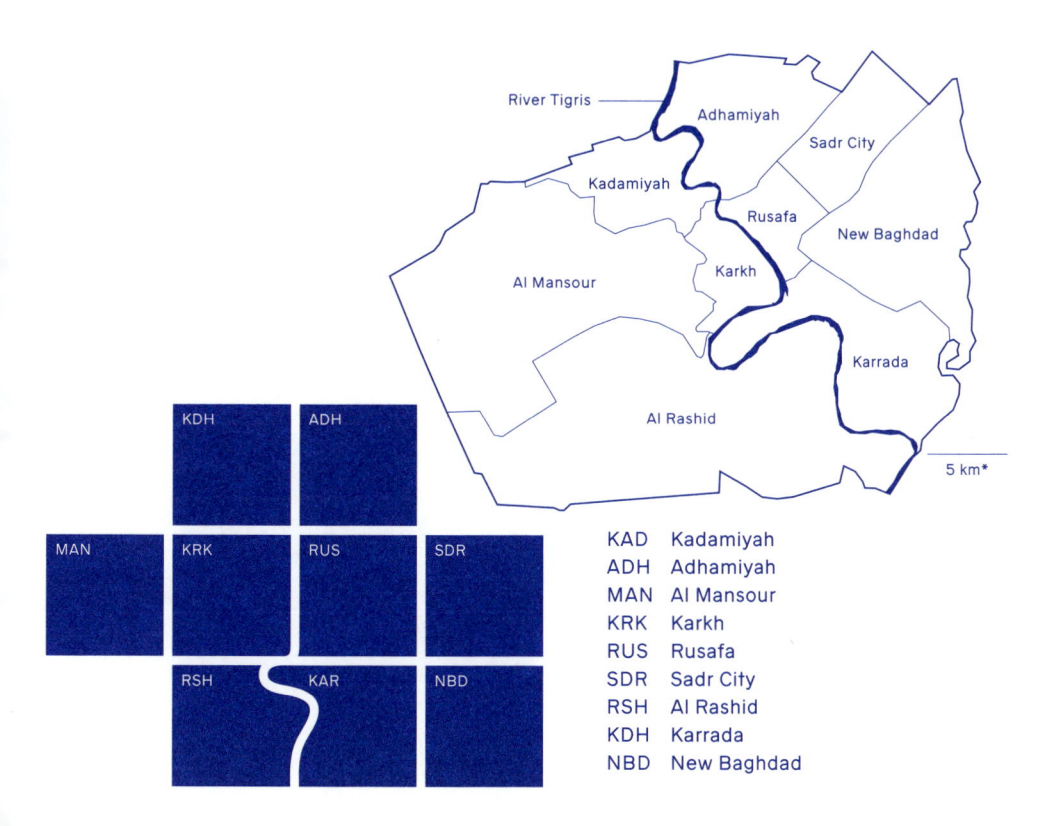

KAD Kadamiyah
ADH Adhamiyah
MAN Al Mansour
KRK Karkh
RUS Rusafa
SDR Sadr City
RSH Al Rashid
KDH Karrada
NBD New Baghdad

720 km²

The area of Baghdad is approximately 720 km²†. Population estimates range from 6,719,500 to 9,760,000 people. The city has nine administrative districts, each of which corresponds to a district advisory councils. The nine administrative districts are subdivided into a further 89 neighbourhoods. (†Many websites, including Wikipedia, list the area of Baghdad as 204.2 km², however this figure does not square with any map we can find. Our figure is derived by calculating the area created by the district map above when it is placed on a geographic map.)

59

SEOUL

As the Thames divides London into two halves, Seoul is divided by the winding path of the River Han. The shape is unique, with an increasingly pronounced curve as it crosses the city. With the Seoul square map, it was challenging to arrange the squares in a manner that closely resembles the real-world arrangement, whilst avoiding gaps. Much experimentation was required. Some liberty was taken with where Seochu-gu was placed in the square map: we placed it caddy-corner to the river but only touching it due to the addition of a curve on the lower right corner of Yongsan-gu.

도봉區	Dobong-gu
노원區	Nowon-gu
강북區	Gangbuk-gu
성북區	Seongbuk-gu
동대문區	Dongdaemun-gu
중랑區	Jungnang-gu
은평區	Eunpyeong-gu
종로區	Jongno-gu
서대문區	Seodaemun-gu
中區	Jung-gu
성동區	Seongdong-gu
광진區	Gwangjin-gu
마포區	Mapo-gu
용산區	Yongsan-gu
강서區	Gangseo-gu
양천區	Yangcheon-gu
영등포區	Yeongdeungpo-gu
동작區	Dongjak-gu
서초區	Seocho-gu
강남區	Gangnam-gu
송파區	Songpa-gu
강동區	Gangdong-gu
구로區	Guro-gu
금천區	Geumcheon-gu
관악區	Gwanak-gu

The area of Seoul is of 605 km², with a population of 9,838,892 people. Seoul is divided into 25 districts (*gu*). The districts vary greatly in area (from 10 to 47 km²) and population (from fewer than 140,000 to 630,000).

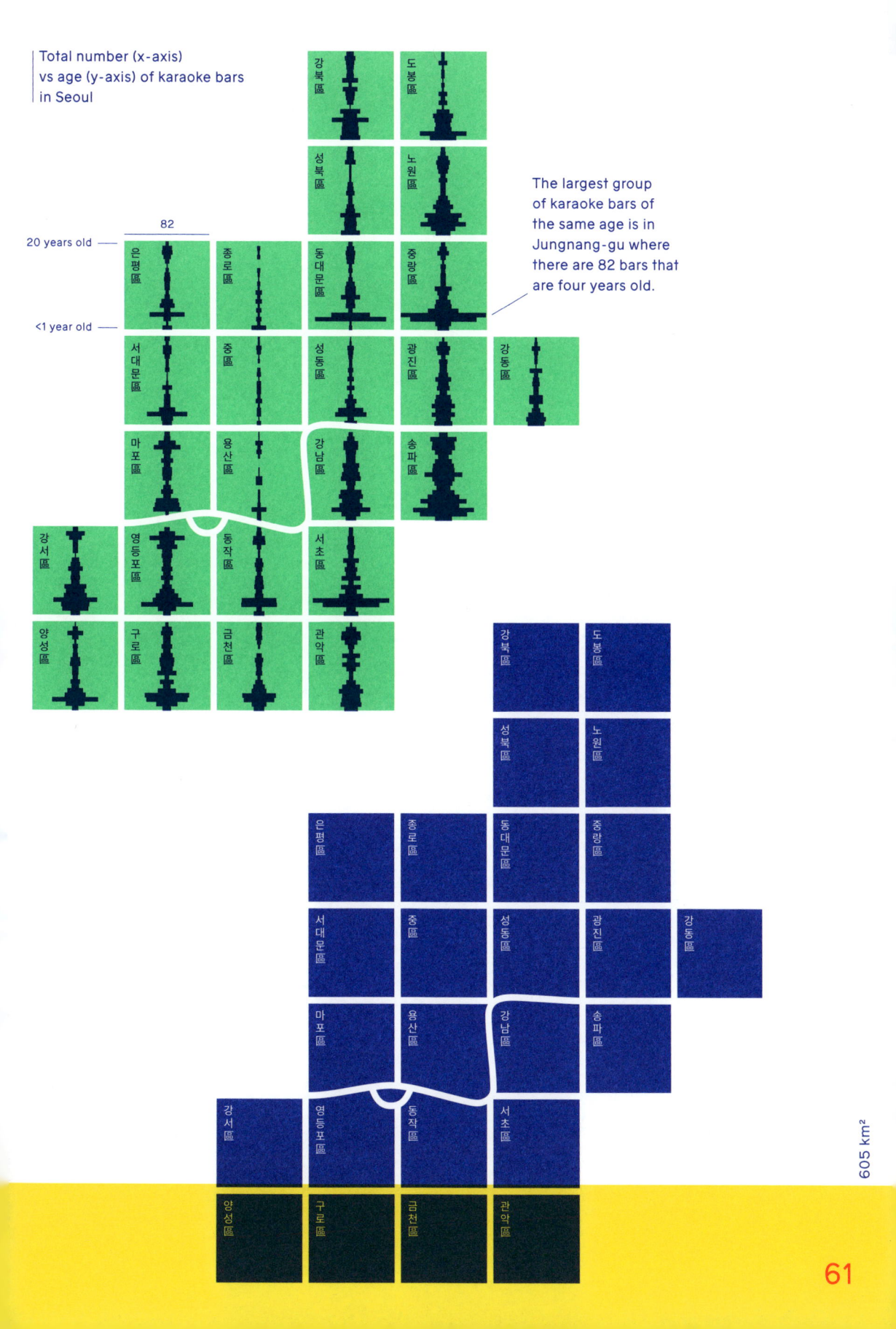

Total number (x-axis) vs age (y-axis) of karaoke bars in Seoul

20 years old

<1 year old

82

The largest group of karaoke bars of the same age is in Jungnang-gu where there are 82 bars that are four years old.

605 km²

61

About map labels   Whenever possible we have labelled the city districts on the geographic maps with their full names, however in many cases this was not possible due to a lack of space. In those instances we have used the same abbreviation as is found in the key for each map.

Sources (by page)

	9	https://en.wikipedia.org/wiki/2014_Scottish_independence_referendum
	10	https://www.ft.com/content/3685bf9e-a4cc-11e6-8b69-02899e8bd9d1

	11	https://commons.wikimedia.org/wiki/File:Mercator_with_Tissot%27s_Indicatrices_of_Distortion.svg
		* Map created by Justin Kunimune (Wikimedia Commons user: Justinkunimune)
		   Creative Commons Attribution-Share Alike 4.0 International license
		
		https://en.wikipedia.org/wiki/Gall%E2%80%93Peters_projection#/media/File:Tissot_indicatrix_world_map_Gall-Peters_equal-area_proj.svg
		* Map created by Eric Gaba (Wikimedia Commons user: Sting)
		   Creative Commons Attribution-Share Alike 4.0 International license

	14	Waldo Tobler, "Thirty-Five Years of Computer Cartograms", in *Annals of the Association of American Geographers,* vol. 94, no. 1, March 2004, pp. 58–73.
	15	Daniel Dorling, *Area Cartograms: Their Use and Creation,* Environmental Publications, 1996. Michael Gastner and MEJ Newman, "Diffusion-based Method for Producing Density-equalizing Maps", in *The National Academy of Sciences of the USA,* vol. 101, 2004, pp. 7499–7504.
	32	https://data.london.gov.uk/dataset/green-and-blue-cover
	33	https://data.london.gov.uk/dataset/historic-census-population https://data.london.gov.uk/dataset/ons-mid-2012-population-estimates
	37	https://theopendata.com/site/2012/01/mumbai-human-development-report-2009/
	39	http://insideairbnb.com/get-the-data.html
	41	https://hri.fi/data/dataset/helsingin-tilastollinen-vuosikirja
	46	http://dati.comune.milano.it/dataset/ds499_open_wifi_2019_distribuzione_per_zona_del_traffico
	49	https://djbooth.net/features/2016-01-05-new-york-city-platinum-albums
	51	http://www.toukei.metro.tokyo.jp/tnenkan/2017/tn17q3e013.htm
	55	https://www.glasgow.gov.uk/index.aspx?articleid=18820
	57	http://tolur.reykjavik.is/PXWeb/pxweb/en/02%20Family%20types/02%20Family%20types__01.%20Nuclear%20families/FJO01002.px/
	61	http://localdata.kr/devcenter/dataDown.do (note: a permalink for the individual dataset is not available)